SpringerBriefs in Electrical and Computer Engineering

W0235021

More information about this series at http://www.springer.com/series/10059

Naser Pour Aryan

Design and Modeling
of Inductors, Capacitors
and Coplanar Waveguides
at Tens of GHz Frequencies

 Springer

Naser Pour Aryan
Institut für Mikroelektronik
Universität Ulm
Ulm
Germany

ISSN 2191-8112   ISSN 2191-8120   (electronic)
ISBN 978-3-319-10186-6   ISBN 978-3-319-10187-3   (eBook)
DOI 10.1007/978-3-319-10187-3

Library of Congress Control Number: 2014947393

Springer Cham Heidelberg New York Dordrecht London

Printed on acid-free paper

Springer is part of Springer Science+Business Media (www.springer.com)

Preface

Extremely and super high radio frequencies (RF) correspond to the ranges of 3–30 and 30–300 GHz, respectively. In these frequency ranges, applications like point-to-point communications and radars have emerged in recent years. RF microelectromechanical systems (RF MEMS) is one available technology to implement circuits in such high frequencies, and is nowadays an interesting research topic. Other more conventional technologies include bipolar junction transistors (BJT), in particular, the heterojunction bipolar transistors (HBT).

Regardless of the used technology, passive components are always necessary for the functionality of the implemented circuit. This book handles the design, modeling, and optimization of these components. Active components like MEMS switches are not investigated. The elements handled involve inductors, coplanar lines, and MIM-capacitors.

In order to study the issue as closely as possible to practice, the components were studied using a definite fabrication technology. The so-called ISiT technology, provided by the Fraunhofer Institute for Silicon Technology (Itzehoe, Germany) is investigated. In addition to the electromagnetic (EM) simulations, some structures were fabricated and measured using the ISiT fabrication process. The research project accomplished here was part of a larger one, mainly corresponding to the application of MEMS structures to frequencies between 24 and 77 GHz. It was called the RF Platform project.

In Chap. 1, the ISiT technology is explained briefly. It is also explained how introducing a polycrystalline silicon layer suppresses the substrate losses, which are usually caused by the inversion channels created under the oxide layer. It is shown how this means that *Agilent ADS* (Advanced Design System, an RF design environment including an EM simulation software called "Momentum") is able to predict the losses of the passive elements, and therefore making loss models for the passive components based on the EM simulations possible.

In Chap. 2 several inductors are designed for operation at frequencies of 24 and 35 GHz. The importance of these two frequencies is due to the specific application in the investigated RF Platform project, which includes these operating frequencies.

The figure of merit for optimizing these inductors was the quality factor. In the design process, it is assumed that the influence of the under-oxide inversion channels is eradicated. Therefore, the substrate was considered as lossless in the EM simulations.

In Chap. 3 a new coplanar line model for the ISiT technology is developed. As this technology has a passivation oxide layer on the substrate, the common coplanar line model available in the literature is no longer applicable. It is shown that the model is able to predict all the necessary parameters of coplanar lines (loss coefficient, characteristic impedance, and effective dielectric constant) on ISiT substrate.

In Chap. 4 a new model for DC-block (MIM-capacitor) was developed. The model is based on basic electromagnetic theory and in part on the line model already presented in Chap. 3. It predicts the simulations (*Sonnet* EM simulations) very accurately.

In Chap. 5 the development of a design-kit for the *Agilent ADS* software is explored. Several pictures are included to show the functionality and different parts of the design-kit. In addition to the DC-block, coplanar line, and the inductors the design-kit includes MEMS switches designed with ISiT. In this design-kit, all the components have an automatic layout generator associated with them and all, except the coplanar lines, can be used for S-parameter simulation. The codes generating the design-kit are presented in the Appendix.

I would like to thank very warmly Prof. Dr.-Ing. Hermann Schumacher and Dr.-Ing. Tatyana Purtova for their great help in accomplishing the current research. Dr. Purtova was always full of innovative ideas and has donated a lot to this work. I also want to thank a lot Mr. Till Feger, especially for his great help in the design-kit development. Finally, I thank my sister Nasim Pour Aryan for motivating me to write this book.

Ulm, July 2014   Naser Pour Aryan

Contents

About the Author

Naser Pour Aryan received the B.E. degree in Electrical Engineering from Ferdowsi University of Mashhad, Iran, in 2005 and the M.Sc. degree in Communication Technology from the University of Ulm, Germany, in 2008. In 2007, he was working in the Institute of Electron Devices and Circuits at the University of Ulm developing passive high frequency elements. He received his Ph.D. degree from the University of Ulm in December 2013. He has been working as a researcher on the retinal prosthesis project at the University of Ulm since January 2008. As part of his tasks, he has been developing on-chip inductors and capacitors for medical applications.

Dr. Pour Aryan's research interests include mainly modeling passive electrical structures, analog and mixed signal circuits, image processing, optics, and electrochemistry.

Acronyms

ADS	Advanced Design System
BJT	Bipolar junction transistors
CPW	Coplanar waveguide
EM	Electromagnetic
HBT	Heterojunction bipolar transistors
MEMS	Microelectromechanical systems
RF	Radio frequency

Chapter 1
RF MEMS Process of Fraunhofer ISiT

In technologies used for high frequencies, high resistivity silicon is usually employed as the substrate material. Its high resistivity reduces the microwave losses. In order to prevent DC-currents and bias-dependent leakage from flowing into the substrate, the silicon surface is typically covered with 500 to 2,000 nm thick layer of oxide. The substrate in the ISiT technology is composed of a 508 μm thick high-resistivity silicon layer with a resistivity larger than 3,000 $\Omega \times$ cm, and a 2,000 nm thermally fabricated silicon oxide layer on top. The available layers in the technology as shown in Fig. 1.1. are as follows (in front you see also their conventional technology names):

1. Under-pass (400 nm thick; material: Ti/Pt/Au/Pt stack) = UPATH
2. Dielectric layers (300 nm AlN, 300 nm SiN) = respectively PASS 2 and PASS 1
3. Sacrificial Cu layer (3 μm) = OS 1
4. MEMS membrane (900 nm Au/Ni/Au) = BRIDGE
5. Gold lines (3 μm) = LINES
6. Thick nickel (15 μm) = SPRING

1.1 Handling the Problem of Inversion Channels

It was shown in [2, 11, 12] that when a structure in which a dielectric oxide layer is present between the bulk high-resistivity silicon of the substrate and the metal layer, the losses are more compared to a structure in which the metal is in direct contact with the silicon. This is due to an inversion region that is built up under the oxide layer due to the surface metal potential, i.e. an induced electron density at the Si/SiO$_2$ interface. One possibility to reduce these losses is the introduction of a polycrystalline silicon layer on the silicon surface [2]. The layer is fabricated using silane based low pressure chemical vapor deposition (LPCVD) and can be 500 nm thick. It is nominally undoped (intrinsic). The polycrystalline silicon layer effectively removes, through traps, any free electrons or holes that may have been induced at the oxide-silicon interface, thus reducing the losses. Another possibility of creating

© The Author(s) 2015
N. Pour Aryan, *Design and Modeling of Inductors, Capacitors and Coplanar Waveguides at Tens of GHz Frequencies*, SpringerBriefs in Electrical and Computer Engineering, DOI 10.1007/978-3-319-10187-3_1

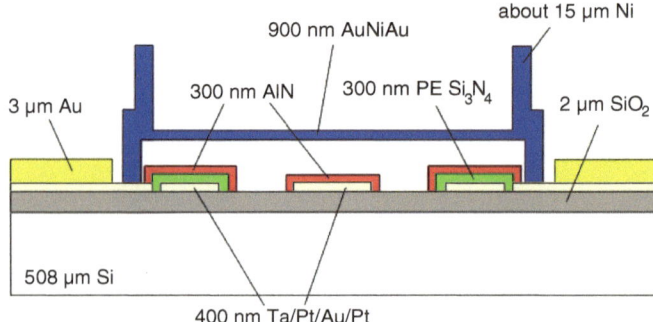

Fig. 1.1 Layers used in ISiT technology

traps is implantation of Argon atoms on silicon surface, for example with a density of $10^{15}\,\mathrm{cm}^{-2}$.

The same method has been used by a new method provided by Fraunhofer Institute to reduce the losses. In order to investigate whether this method is helpful and reduces the losses, two sets of coplanar lines, with similar geometrical and physical characteristics, but different substrates were measured. The difference in the substrate is that in one set of these coplanar lines a poly-silicon layer has been used, but in the other the substrate is the normal ISiT substrate without poly-silicon layer.

1.1.1 Measurement Results

The layout of one of the measured coplanar lines is shown in Fig. 1.2. The coplanar line is built of the gold layer of the ISiT technology.

In Fig. 1.2 the round red dots show the position of the probes in the measurements. It was not possible to place the probes right on the edges of the lines due to a limited probe pitch (max. 150 μm). As a result, the matching at higher frequencies is deteriorated considerably. Figure 1.3 shows the measured forward voltage gain S_{21} in dB. The line on the substrate with a poly-silicon layer has higher S_{21} and therefore lower losses. The fast degradation of S_{21} at higher frequencies is due to the increase of S_{11}. To evaluate actual losses, the loss factor defined as

Fig. 1.2 The coplanar line measured on two different substrates, layout shown in ADS

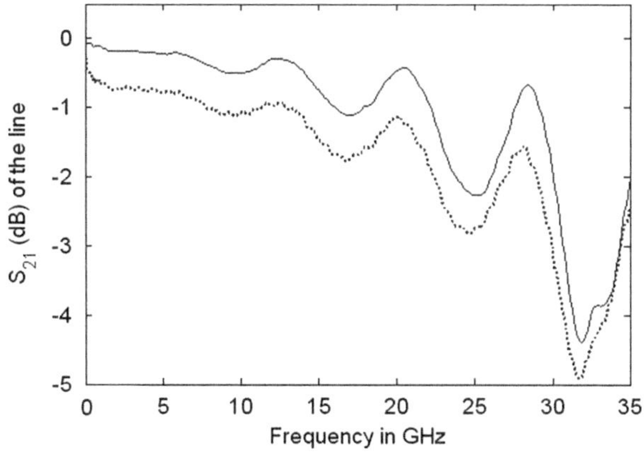

Fig. 1.3 Measured S_{21}, **a** for coplanar line on the substrate with poly-silicon (*solid line*), **b** for coplanar line on the substrate without poly-silicon (*dotted line*), as it is seen here S_{21} is much higher for the coplanar line on the substrate with poly-silicon, thus the loss is lower for this. The removed channels might be a reason for it

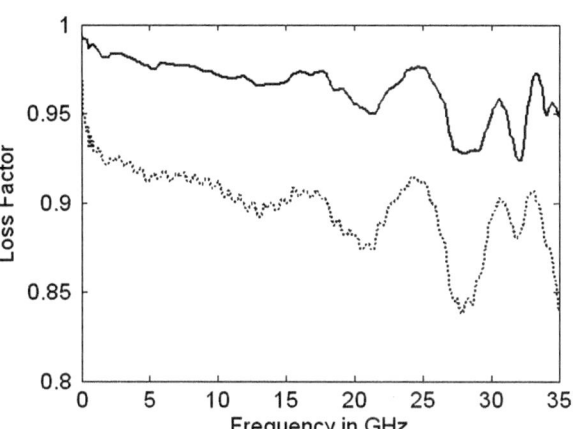

Fig. 1.4 Measured loss factor, **a** for coplanar line on the substrate with poly-silicon (*solid line*), **b** for coplanar line on the substrate without poly-silicon (*dotted line*), as it is seen here loss factor is much higher for the coplanar line on the substrate with poly-silicon, thus the loss is lower for it

$$LF = \sqrt{|S_{11}|^2 + |S_{21}|^2} \tag{1.1}$$

is used and is plotted (Fig. 1.4). As you can see the loss for the structure with poly-silicon is much lower (the higher the LF, the lower the losses). Therefore, the positive influence of the poly-silicon layer in removing inversion channels under the oxide layer is confirmed.

Assuming an "$L_S R_S G_S C_S$" model for transmission lines (Fig. 1.5), the shunt resistance to the ground ($1/G_S$) for both lines was extracted (Fig. 1.6). The line

Fig. 1.5 Transmission line model

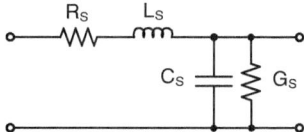

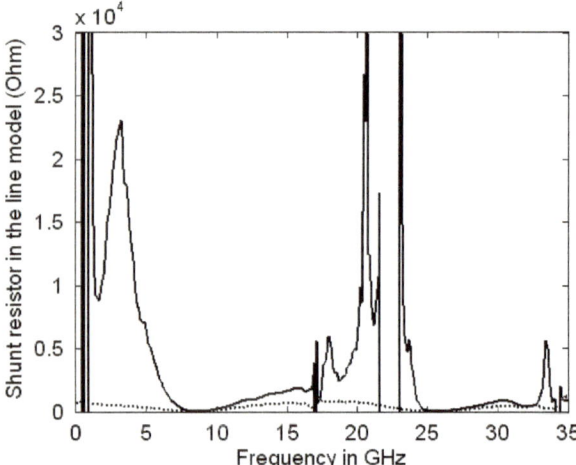

Fig. 1.6 Shunt resistor (inverse of G_S) in the line model for **a** for coplanar line on the substrate with poly-silicon (*solid line*), **b** for coplanar line on the substrate without poly-silicon (*dotted line*), as it is seen here the shunt resistor is much higher for the substrate with poly-silicon. This is an indication that the inversion channels are eliminated

corresponding to the substrate with poly-silicon has a much higher shunt resistance $1/G_2$. This is another indication that the inversion channels are not present any more.

1.1.2 Simulating with ADS Software

Simulations were run with ADS using the built-in model for coplanar waveguide (CPW) lines. The geometrical characteristics of the above line were used as the parameters of this model. The circuit used for the simulation is shown in Fig. 1.7.

ADS simulation ports are placed at the location of probes during the measurement. The two pairs of the coplanar lines on the two sides of the arrangement of Fig. 1.7 are used to model the discontinuities on the two ends of the coplanar line structure. Simulation and measurement results for S_{21} and loss factor are shown in Fig. 1.8a, b for the coplanar line on substrates with and without poly-silicon. In these figures the

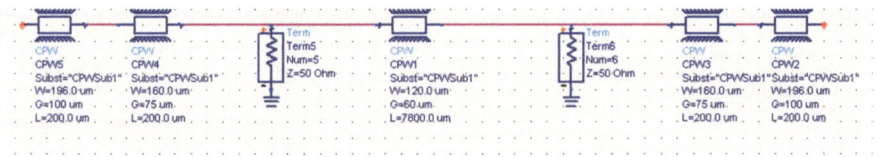

Fig. 1.7 ADS simulation setup for the coplanar line of Fig. 1.2

Fig. 1.8 Above S_{21} and below loss factor, **a** for the coplanar line measured on the substrate with poly-silicon (*solid line*), **b** for the coplanar line measured on the substrate without poly-silicon (*dotted line*), **c** for coplanar line simulated by ADS having no substrate losses (*dashed line*), the agreement between the ADS loss assessment and the measurement result for losses for the coplanar line on a substrate with poly-silicon is acceptable. Therefore, for the passive elements, the EM simulator (for example ADS Momentum) is used for modeling the losses while setting the substrate lossless

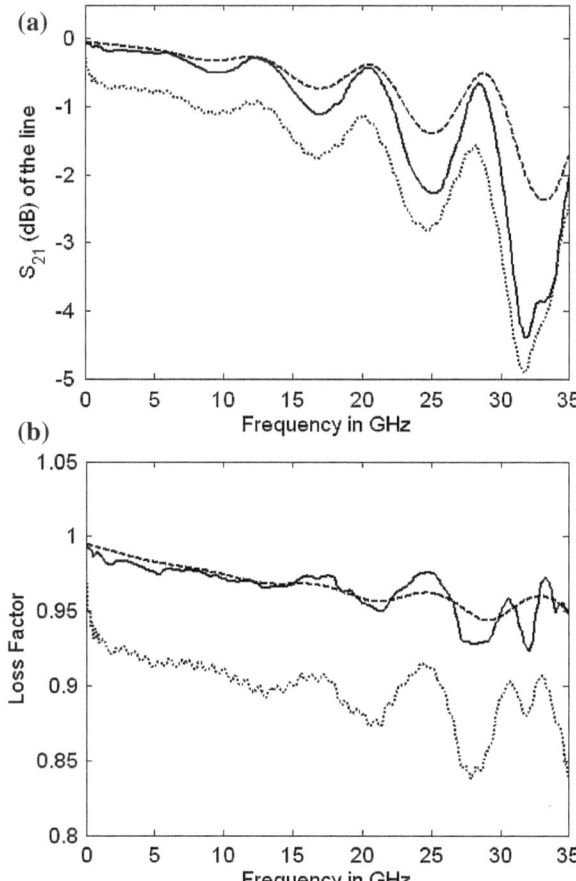

simulation results of the ADS model for a coplanar line fit very well to the measured graphs of the line on the substrate with poly-silicon. In the component model for the coplanar line in ADS the substrate losses were set to zero, so no inversion channel is available in the fabricated structure any more.

The above conclusion is important for the modeling and design procedure developed in the following chapters. The substrate is assumed constantly as lossless.

Chapter 2
Designing Inductors

Spiral inductors are commonly used in radio-frequency integrated circuits (RFICs) to act as series or shunt elements in the matching, tank or choke circuits. The quality factor (Q) is among the most important parameters in order to evaluate an inductor's performance. Unfortunately, quality factor and frequency limitations limit RF front-end circuitry to a large number of discrete passive components and make RF front-end module integration difficult. High-quality-factor inductor design and fabrication remains a challenge for applications that depend on passive components performance, e.g. low phase-noise voltage-controlled oscillator (VCO), power amplifier (PA), low noise amplifier (LNA) and double-balanced Gilbert-cell mixers.

Here two libraries of inductors have been designed. The libraries were designed for 24 and 35 GHz. These two along with 77 GHz are among the frequencies of interest in the RF Platform project. The inductors were designed to achieve a quality factor as high as possible at these frequencies using the ISiT technology, while at the same time occupying the smallest possible area. In total 8 inductors were designed.

In order to enhance the quality factor of the inductors, a suspended structure was used. In all of the structures there is $3\,\mu$m distance between the spiral part and the surface of the substrate. This is done by conventional MEMS micro-machining, i.e. electroplating of the structures and removal of the sacrificial layer. Later the measurements of the fabricated inductors revealed that this suspension would not have a considerable effect if the inversion layer on the silicon surface is not eliminated using poly-silicon or Argon traps. This will be presented later.

The sacrificial layer material used is copper which is named OS1 (German: "Opfer Schicht 1" meaning sacrificial layer (1) in the ISiT technology.

2.1 The Underlying Theory

The optimization of RF inductor performance requires the identification of the relevant parasitic components and their effects. Physical modeling leads to in-depth understanding of the devices [13].

© The Author(s) 2015
N. Pour Aryan, *Design and Modeling of Inductors, Capacitors and Coplanar Waveguides at Tens of GHz Frequencies*, SpringerBriefs in Electrical and Computer Engineering, DOI 10.1007/978-3-319-10187-3_2

A rough model that will be discussed throughout the optimization procedure is depicted in Fig. 2.1. L_S and R_S are the inductance and the conductor resistance. C_S models the capacitive coupling between neighboring turns. C_{SUB1} and C_{SUB2} model the capacitive coupling between the substrate and the ports. The substrate losses are neglected here.

With the help of the π-model of Fig. 2.2 which includes the Y-parameters (admittance parameters) the model elements can be calculated. The parameters L_S, R_S, and C_{SUB1} and C_{SUB2} are determined.

The impedance of C_S is high at the frequencies under the self-resonance frequency of the inductor and is thus negligible in the parallel connection. For L_S and R_S:

$$- Y_{21} = \frac{1}{R_S + j\omega L_S} \Leftrightarrow R_S + j\omega L_S = -\frac{1}{Y_{21}} \tag{2.1}$$

Therefore:

$$R_S = -Real(\frac{1}{Y_{21}}) \tag{2.2}$$

and

$$L_S = -\frac{1}{\omega} \cdot Imag(\frac{1}{Y_{21}}) \tag{2.3}$$

Fig. 2.1 The simple inductor model, substrate losses are neglected

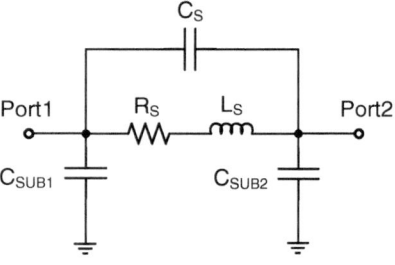

Fig. 2.2 π-model consisting of the Y-parameters

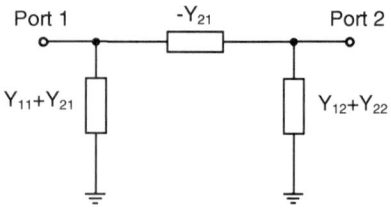

For C_{SUB1}:

$$Y_{11} + Y_{21} = j\omega C_{SUB1} \tag{2.4}$$

Therefore:

$$C_{SUB1} = Imag(\frac{Y_{11} + Y_{21}}{\omega}) \tag{2.5}$$

With the same method C_{SUB2} can be calculated:

$$C_{SUB2} = Imag(\frac{Y_{12} + Y_{22}}{\omega}) \tag{2.6}$$

A more complete model that can also explain the decrease of the measured series resistance of the inductor with rising frequency despite of the skin effect and eddy currents is seen in Fig. 2.3 [8]. It arises when the inductors are built on a substrate which has an additional oxide layer on top, as is the case in the ISiT technology. While it is expected that the measured series resistance of the inductor increases with frequency, because of the effect of the capacitor C_{EL}, it decreases in reality [8]. C_{EL} models the capacitive coupling between the neighboring turns through the silicon substrate. C_{OX1} and C_{OX2} model the coupling between the inductor and the substrate. It can be shown that the presence of C_{EL} causes the measured series resistance of the inductor to begin to decrease at frequencies around the peak of the quality factor of the inductor.

However, in this essay the first model is considered as valid enough.

In order to have a high resonance frequency and thus a high operation frequency (i.e. frequency of maximum Q), the parasitic capacitance C_S which resonates with the inductor (in Fig. 2.1, parallel to the series connection of the resistor and the inductor) must have values as small as possible. This parasitic capacitor can be calculated from the resonance frequency (f_0) of the inductor with the following formula:

$$C_S = \frac{1}{L_S \cdot (2\pi f_0)^2} \tag{2.7}$$

Fig. 2.3 The more complete model for inductor (substrate losses are neglected)

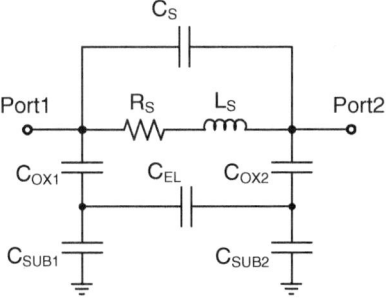

Fig. 2.4 The 3-dimensional view of one of the designed inductors and its cross-section in the AA' plane (not to scale)

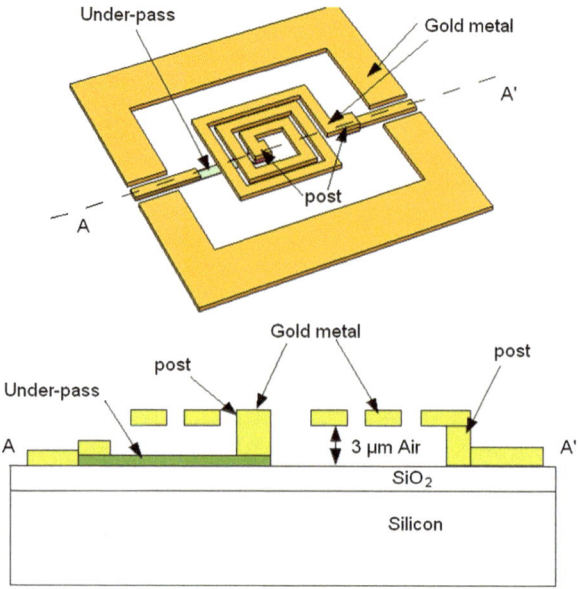

The parasitic capacitor C_S can be calculated from mathematical formulas available to calculate the capacitance between parallel stripes of metal. Figure 2.4 shows the structure of one of the inductors. The inductor is made of several straight pieces of gold metal. It is possible to consider the turns of the inductor as pieces of microstrip lines that have capacitive coupling to each other, and calculate the "per length" capacitance between the two microstrip lines with the same geometrical characteristics and the same distance as the inductor turns, and then multiply the result by the total length on which the two turns neighbor each other. One method to calculate this capacitance between coupled microstrip lines is described in [9]. The parasitic capacitance from this method matches the values calculated from Eq. 2.7 for different inductors quite well. The calculated values for C_S are quite small, and for the structures designed here they range from 20 to 50 fF.

Therefore, one major goal in the design strategy would be designing inductors with trivial C_S values. For example, a long length of adjacent turns and too wide metal stripes must be avoided (wide stripes result in a lot of capacitive coupling).

As an attempt to have low parasitic capacitance, it was tried to minimize it by using a narrower metal stripe for the under-pass, which connects the termination of the inner turn of the inductor to the outside with a metal layer that is directly on the substrate (the whole structure is in the air). These two metal layers, i.e. the structure in the air and the under-pass, are directly on top of one another (as is seen in Fig. 2.4), so making the under-pass narrower would potentially reduce the parasitic capacitance. However, simulations showed that the structure with a narrower under-pass does not have a considerably higher resonance frequency. Consequently, this was ignored in the design.

Capacitor C_S, albeit being small, is large enough to prevent using spiral inductors for 77 GHz. So according to the simulations it is not possible to use spiral inductors here. The only way to have a passive inductor in ISiT technology at this frequency is to use a transmission line.

2.2 Accurate Formula for the Q-factor in Two-Port Configuration

Usually in the literature the Q-factor is formulated rigorously as the ratio of imaginary part to real part of input impedance of the inductor in a one port configuration while the other port is shorted to ground [5]:

$$Q = \frac{Imag\{Z_{in}\}}{Real\{Z_{in}\}} \qquad (2.8)$$

This Q-factor formulation is proper for spiral inductors when serving as shunt elements. However, spiral inductors are also used frequently as series elements in many applications and their performance is lower than anticipated by the one-port formulation in these cases. The formulas required in these conditions are usually very large [5], but when the reflection coefficients (Γ) at the two inductor ports are zero meaning that the inductor is completely matched to the transmission lines connected to it, the following short expression can be used (S are the scattering parameters):

$$Q = \frac{2Imag(S_{11})}{1 - |S_{11}|^2 - |S_{21}|^2} \qquad (2.9)$$

2.3 Labeling Method of the Inductor Structures

In the following for designating the inductors special labels are used. Let's explain this by an example. Consider the following label:

1.5_120_20_10

This means that the corresponding inductor has one and a half turns, has an inner diameter of 120 μm, a conductor metal width of 20 μm and a turn-to-turn spacing of 10 μm.

2.4 Design Parameters

There are four important parameters affecting the inductors' performance [10]:

2.4.1 Inner Diameter

Inner diameter has a significant effect on the quality and the area consumption of the inductor [10].

There is an optimum inner diameter at which the Q-factor attains higher values [10]. This depends on the technology being used. In [10] for example, it was $100\,\mu$m. For diameters less than this value, Q will be smaller at the operating frequency. The reason: When the inner diameter is small, the majority of the changing magnetic flux due to the current in outer turns passes through the metal conductor of the inner turns, inducing eddy currents there. These eddy currents end up in increased losses in the inductor, lowering the Q-factor. Increasing the inner diameter causes this changing magnetic flux to pass through the empty area in the middle of the inductor instead, thus decreasing the losses.

As the technology and material were different here, especially that here high resistivity silicon and in [10] low resistivity silicon was used, simulations were necessary to extract this optimum inner diameter. It proved that in ISiT technology this diameter amounts to $120\,\mu$m.

Many simulations were run to verify that $120\,\mu$m is the best inner diameter. As an example the results of simulations for several different inductors are shown in Table 2.1. The Q-factor curves for 2_60_20_10, 1.5_120_20_10 and 1.5_80_20_10 are seen in Fig. 2.5. As it is seen in this figure, the Q-factor curve versus frequency increases with increasing inner diameter, so it seems reasonable to increase this parameter up to a value of $120\,\mu$m to acquire a better performance for the inductor.

As it is visible in Table 2.1, with increasing the diameter of the spiral inductor up to $120\,\mu$m, the Q-factor and the inductance value increase simultaneously. However from $120\,\mu$m on, although the inductance increases, Q-factor starts to decrease due to additional parasitic effects and high losses caused by the long length of the conductor. Therefore, from the point of view of both Q-factor and inductance, $120\,\mu$m is the optimal inner diameter in ISiT technology. Anyway, other inner diameters are sometimes necessary to achieve various values for the inductance.

Table 2.1 Comparison between inductors having different inner diameters

Inductor label	Q at 24 GHz	L (nH)	Inner diameter (μm)
2_60_20_10	12.9	0.69	60
1.5_80_20_10	14.3	0.65	80
1.5_100_20_10	20	0.8	100
1.5_120_20_10	21.6	0.88	120
1.5_140_20_10	15.9	0.99	140

Fig. 2.5 Q-factor versus
frequency for **a** 2_60_20_10:
inner diameter 60 μm (*solid
line*), **b** 1.5_80_20_10: inner
diameter 80 μm (*dotted line*),
c 1.5_120_20_10: inner
diameter 120 μm (*dashed
line*), as it is seen here up to a
diameter of 120 μm, adding
the inner diameter increases
the quality factor versus
frequency

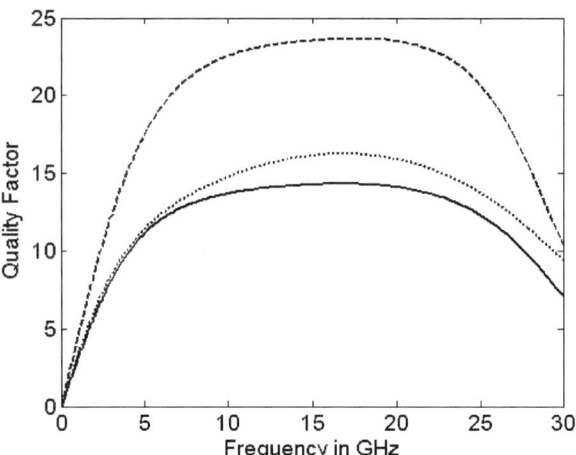

2.4.2 Turn-to-Turn Spacing

The metal conductor-to-conductor spacing does not have a major effect on the quality
factor [10]. With decreasing spacing, on one hand the inductance increases due to a
higher turn-to-turn inductive coupling. However, on the other hand the high frequency
resistance also rises due to proximity effects. The quality factor is determined by the
resulting inductive and resistive behavior of the inductor. Consequently, in theory, the
counter-compensation between the inductance and the resistance results in similar
quality factors for inductors having different turn-to-turn spacing. Since minimum
spacing leads to least chip area consumption and highest inductance, it seems to be
the best option.

The simulations showed that lower spacing, for the same inductance values,
provides a better Q-factor behavior. For example in Fig. 2.6, which compares the
Q-factor of 1.5_140_20_10 with 1.5_120_20_20 over frequency, although they have
the same inductance of 1 nH, the one with smaller spacing has much higher Q values.
Other simulations confirmed that always the inductors with lower spacing are supe-
rior. Consequently, the lowest spacing was almost always used, with one exception.
Among the final inductors, the 0.5 nH inductor at 35 GHz has a larger turn-to-turn
spacing of 20 μm, while the others have a spacing of 10 μm. This inductor is very
small (geometrically) compared to the others. Concerning the fact that the structures
are elevated and as is shown in Fig. 2.4, posts are needed to maintain them in the
air, for a very small inductor like this which has a very small inner diameter, the
two posts on the two sides of the structure would be quite near to each other. In
order to prevent the two posts from causing a large capacitive coupling to each other
and to the inductor body, higher spacing was chosen in this case. The posts were
consequently farther from each other.

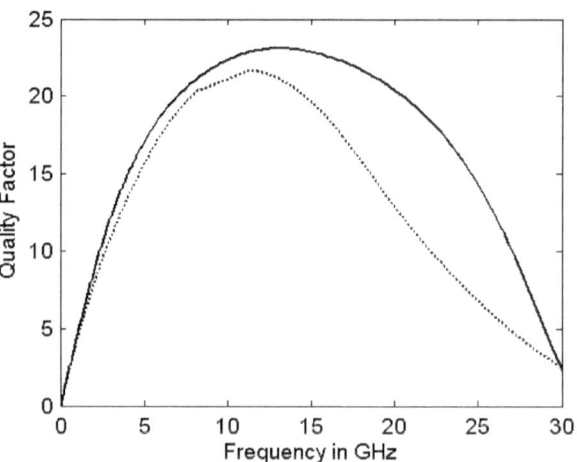

Fig. 2.6 Q-factor versus frequency for two inductors with equal inductance of 1 nH and with turn-to-turn spacing of **a** 10 μm (*solid line*), **b** 20 μm (*dotted line*), as it is shown here, a smaller spacing results in higher quality factor over frequency

2.4.3 Conductor Metal Width

Conductor metal width is at first glance the most complicated parameter of all. A smaller conductor width would mean less substrate and eddy current losses but at the same time higher skin effect losses. From another point of view, more width results in higher turn-to-turn capacitive coupling, and thus higher C_S. An optimum metal width at which the best compromise between these different effects exists must be determined. The simulations show that here, the best width is the least possible width in the technology (20 μm for the LINES layer), giving highest Q-factor values and resonance frequencies. There were no exceptions here. Attention that these results may only hold for this technology. It is possible that in another technology the least possible metal width is not the optimal metal width, because of for example, too high conductor losses due to a too narrow metal width.

In addition to the above parameters, another important parameter to be considered in the inductor design is the used material:

2.4.4 The Material Used

In the ISiT technology there are two metal layers that could be used for building the inductors: thick nickel (SPRING) and gold lines (LINES). The UPATH layer was completely ignored due to its small thickness and high material resistivity. Simulations showed that the inductors made of nickel exhibit much lower Q-factor. Fig. 2.7 illustrates the Q-diagrams of two inductors with equal inductances, one made of nickel and the other made of gold. The inductor made of gold has a much better Q performance. This is due to the higher resistivity of nickel compared to gold. All of

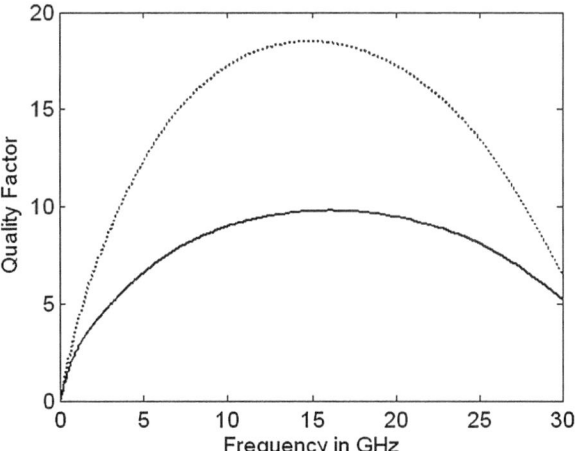

Fig. 2.7 The Q-factor versus frequency for two inductors with equal inductance made of **a** gold (*dotted line*) and **b** nickel (*solid line*), as it can be seen, the gold structure has a much better Q-factor over frequency behavior than the nickel structure

the final inductors, except one, were made of gold. The reason that this one inductor (0.76 nH for 35 GHz) is made of nickel is that the width of nickel metal in the used technology can be made narrower compared to gold (10 μm compared to 20 μm). So it is possible to make a physically smaller inductor having higher inductance with nickel by using a higher number of turns. Smaller inductor would mean less loss and less parallel shunt capacitance, due to less wiring length. Therefore, higher inductance and Q-factor values were attainable. But generally gold is better.

The above discussions also hold for the 35 GHz frequency, so the same strategy has been used to design the inductors for this frequency. In the following the method used to design the inductors is explained.

2.5 The Strategy for Inductor Library Design

While at lower frequencies (for example tens of MHz) it is relatively easy to attain an inductor design with an adequately large Q factor, at higher frequencies it is a challenge that the final design would feature a large enough (or at least positive) Q, due to the high effect of the parasitic components, especially the inter-turn capacitor C_S. At tens of MHz frequencies, the design of an inductor for a definite inductance value is possible through the application of the modified Wheeler expression [7]:

$$L_{mw} = K_1 \mu_0 \cdot \frac{n^2 d_{avg}}{1 + K_2 \rho} \tag{2.10}$$

$\mu_0 = 4\pi \times 10^{-7} \frac{H}{m}$ is the vacuum permeability and n is the number of turns. d_{avg} is the arithmetic mean value of the inner and outer diameters as shown for an square inductor in Fig. 2.8:

Fig. 2.8 Square inductor
topology

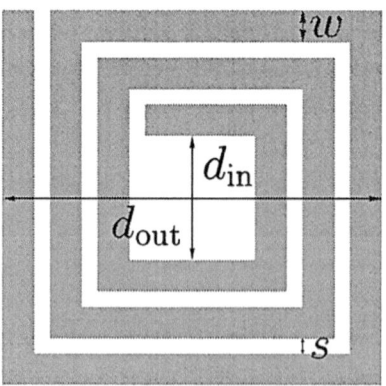

$$d_{avg} = \frac{1}{2} \cdot (d_{out} + d_{in}) \tag{2.11}$$

ρ is the fill ratio which is dimensionless and is calculated by:

$$\rho = \frac{d_{out} - d_{in}}{d_{out} + d_{in}} \tag{2.12}$$

K_1 and K_2 are layout-dependent coefficients. For square shape $K_1 = 2.34$ and $K_2 = 2.75$.

However, for higher frequencies investigated here it was preferred to focus on increasing the Q factor and decreasing the area. The accurate inductance values were extracted at the end, after the design.

For two key parameters, conductor to conductor spacing and conductor metal width, the values were fixed in the previous section through a large number of simulations. Therefore, only two parameters are remaining: Inner diameter and the number of turns. The number of turns used was 1.5 and 2.5 turns. Integer numbers (like 1 and) were not used for the number of turns simply because geometries corresponding to 1.5 and 2.5 turns are easier to draw and build.

Using 3.5 turns is not possible according to the simulations. Such an inductor, even with the least possible inner diameter of 40 μm (thus being small and having a small parasitic parallel capacitance) would already possess a resonance frequency below 24 GHz.

For the smallest inductor (0.3 nH) a small radius was used. However, this inductor does not have a complete turn, otherwise it becomes too large. The next larger inductor (0.53 nH) has 1.5 turns and a diameter of 60 μm. Because having a large inner diameter is advantageous, for the smaller inductors no 2.5 turns has been used, otherwise the inner diameter must be made small to acquire smaller values for the inductance, resulting in bad Q behavior.

As noted above, inner diameters larger than an optimal value will not improve the Q factor anymore. So for higher inductances, instead of having a very large inner diameter which in turn introduces lots of conductor losses and parasitic capacitances (due to the large size of the structure), another turn is added. The inductance value is proportional to the square of the number of turns, so it grows fast with adding more turns. Generally larger inductance values must have more turns rather than a very large inner diameter.

The largest inductors in 24 and 35 GHz frequencies have 2.5 turns.

2.6   Inductor Libraries

Finally two libraries of inductors for 25 and 35 GHz were developed. Pictures of some of these are illustrated later in Chap. 5.

In Table 2.2, the characteristics of the inductors designed for 24 GHz are seen. The label for each structure is visible in the table. Two Q values are available for each inductor. One is calculated from the accurate "two-port" formula for Q (Eq. 2.9) and the other is calculated from the common formulation for the Q (Eq. 2.8) which is more appropriate when the inductor is used as a shunt element.

All these inductors are made of gold. At 24 GHz no nickel was used as the structure material.

In Table 2.3 the characteristics of the inductors designed for 35 GHz are illustrated. The largest inductor has again 2.5 turns and is made of nickel, the only inductor made of nickel in these two libraries. Its value is not much higher than the next inductor in the group which is made of gold but it has a higher Q.

Table 2.2 The characteristics of the inductors designed for 24 GHz operation frequency

Inductor label	L (nH)	Q (two port formula)	Q (common formula)
Piece of line	0.3	10.4	19.2
1.5_60_20_10	0.53	11	16.4
1.5_110_20_10	0.85	11.5	15.4
1.5_140_20_10	1	9.5	12
2.5_70_20_10	1.3	4.2	5

Table 2.3 The characteristics of the inductor library for 35 GHz

Inductor label	L (nH)	Q (two port formula)	Q (common formula)
Piece of line	0.3	9	18.5
1.5_40_20_20	0.5	8	13
1.5_90_20_10	0.7	4	6.2
2.5_40_10_10	0.76	3	4.3

It is visible in the tables that the Q values calculated from the formula for the inductor used in the shunt arrangement is always higher than the accurate two port value.

2.7 Measurement Results

Some of the designed inductors were fabricated using ISiT and measured for their Q factors, both with and without inversion channel suppression method. The results are depicted in Fig. 2.9. At 24 GHz, inductors with values of 0.53, 0.85 and 1.3 nH and at 35 GHz, 0.3, 0.5 and 0.7 nH, were measured. As can be seen from the figure, for the two-port formula and especially at 35 GHz, the inductors fabricated on the substrate having deposited poly-silicon feature larger Q factors than was expected from the simulations (Tables 2.2 and 2.3).

As it is seen in this figure, although the structures are elevated, the interaction between the inductor and the substrate still exists.

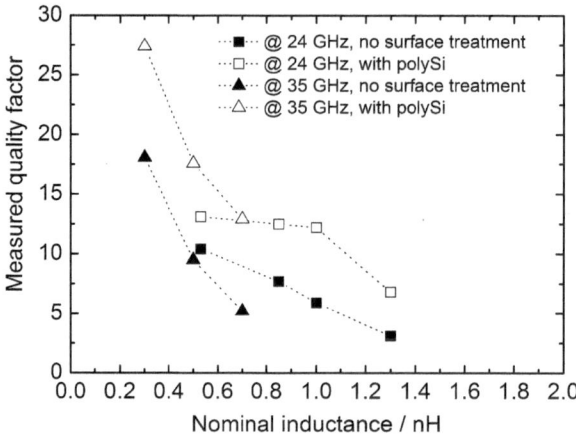

Fig. 2.9 Measured quality factors for the two sets of elevated micromachined spiral inductors, optimized for the 24 and 35 GHz frequencies

Chapter 3
Modeling the Coplanar Line

3.1 The Theoretical Approach

A coplanar line (waveguide) is composed of a signal track with two return conductors on its two sides. All three are in the same plane.

In the literature there are relatively accurate and complete models for coplanar lines which are fabricated on homogeneous substrates. However, in ISiT and many common RF technologies the substrate is covered by a thin layer with a different dielectric constant on top, i.e. a 2 μm thick layer of silicon oxide having a dielectric constant of 4. In the following, a model that is specific to the ISiT technology is developed. This model is valid when the bulk material of the substrate is silicon which has a dielectric constant of 11.9 covered by a 2 μm thick oxide layer with a dielectric constant of 4.

The method used to derive the model is based on the existing formulas for the homogeneous substrate case, together with electromagnetic concepts and Sonnet (a common EM simulator) simulations. The modeling process explained in the following can be applied to other similar substrate structures. A cross section of the ISiT coplanar waveguide structure is shown in Fig. 3.1. C'_{OXSIG} and C'_{OXGND} are the capacitance per unit length between the signal track and the return conductors on one side and the silicon substrate from the other side, respectively.

If the unit length is considered to be very short, a piece of coplanar line with the unit length and lacking any losses will have a model roughly looking like Fig. 3.2. C' and L' are the distributed capacitance and inductance of the line, respectively. Distributed value means value per unit length. The capacitor is due to the capacitance between the signal and the return conductors. The inductor is due to the magnetic fields developed around the signal track.

The presence of a dielectric does not affect the value of L'. This is because dielectrics are not magnetic materials and their relative permeability constant is equal to one, and so their presence does not affect the magnetic flux surrounding the coplanar line. Therefore, for any kind of dielectric arrangement, L' will always remain constant. The parameter which depends on the dielectric constant of the surrounding material is C'.

In the case of a homogeneous substrate, without the backside metal, C' will be equivalent to the parallel combination of two capacitors, one due to the field lines

© The Author(s) 2015
N. Pour Aryan, *Design and Modeling of Inductors, Capacitors and Coplanar Waveguides at Tens of GHz Frequencies*, SpringerBriefs in Electrical and Computer Engineering, DOI 10.1007/978-3-319-10187-3_3

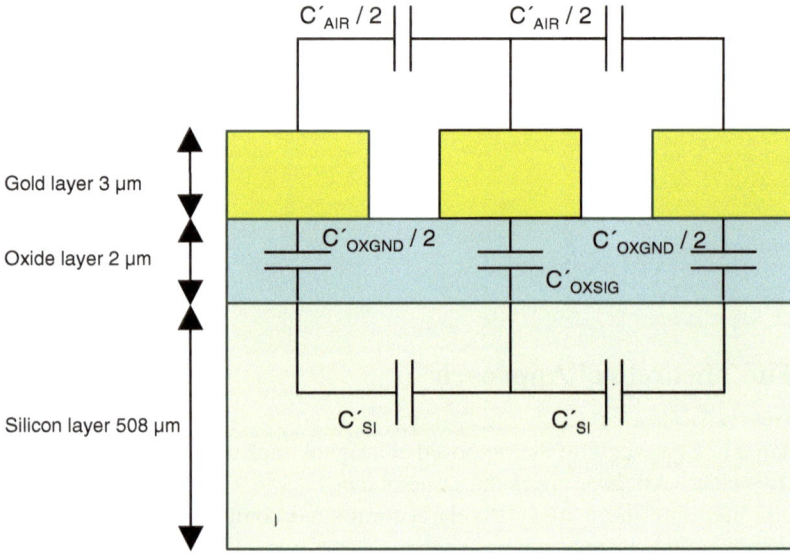

Fig. 3.1 The cross section of the coplanar *line*, the necessary capacitors for the modeling of the ISiT coplanar waveguide are shown

Fig. 3.2 The general model of a short piece of coplanar *line*

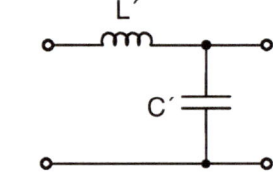

Fig. 3.3 The model of the *line* on homogeneous substrate

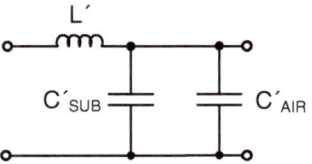

that penetrate through the substrate, and the other due to field lines that flow through the air. The model for this case (homogeneous substrate) is depicted in Fig. 3.3. Note that this is the same as the above mentioned general model, while the different constituents of C' are simply designated separately.

A method to calculate the capacitors C'_{SUB} and C'_{AIR} for metals with zero thickness is mentioned in [3]. This method is a quasi-static method and is based on a sequence of conformal mappings. It holds for frequencies as high as 40 GHz. Here it has been partially modified and is explained as follows:

In both cases (capacitive coupling through the air and capacitive coupling through the substrate), the arrangement is similar to a metal stripe between two other metal stripes (return ground plates here). This arrangement is shown in Fig. 3.4a. Figure 3.4 contains four planes with their corresponding two dimensional variables (z, t, x and w respectively). z plane is the plane where the structure of the coplanar line lies. w plane is the plane in which the equivalent parallel plate structure to the coplanar line configuration after the mappings lies, and t and x planes are the planes of the intermediate geometrical configurations in the mappings, between the z plane and the w plane.

For calculating the capacitance through the air the first quadrant of Fig. 3.4a is transformed into the upper t half-plane of Fig. 3.4b by the mapping $t = z^2$ and then into the rectangular region of Fig. 3.4d through the following geometrical mapping:

Fig. 3.4 Conformal mapping for coplanar waveguide:
a original structure;
b intermediate transformed plane for the *dashed* region;
c intermediate transformed plane for the *dotted* region;
d final mapping into a plane-parallel capacitor (valid for both regions), picture from [3], with permission

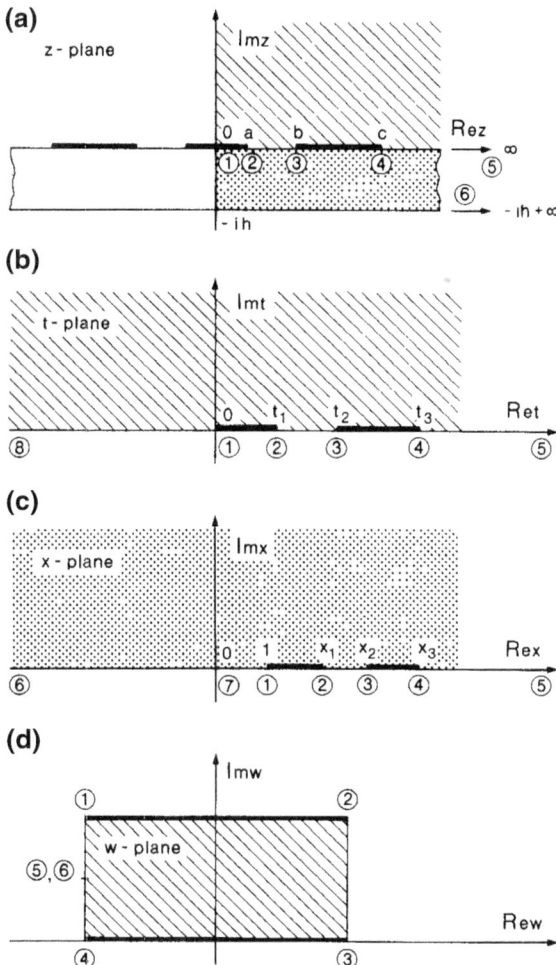

$$w = \int_{t_0}^{t} \frac{dt}{\sqrt{t \cdot (t-1) \cdot (t-t_1) \cdot (t-t_2)}}$$

The distributed capacitance through the air will be equal to:

$$C_a' = 2 \cdot \varepsilon_0 \cdot \frac{d_{12}}{d_{23}} = 2 \cdot \varepsilon_0 \cdot \frac{K(k_1)}{K(k_1')} \tag{3.1}$$

Here $K(k_1)$ and $K(k_1')$ are the complete elliptic integrals of the first kind, $k_1' = \sqrt{1 - k_1^2}$ and

$$k_1 = \frac{a}{b} \cdot \sqrt{\frac{1 - \frac{b^2}{c^2}}{1 - \frac{a^2}{c^2}}}$$

Finally d_{12} and d_{23} are the distances between points 1 and 2 and points 2 and 3 in the w plane, respectively. In the above formula a, b and c are the geometrical parameters of the coplanar structure as shown in Fig. 3.4a.

The capacitance through the substrate is more complicated to assess. This is calculated by summing the air contribution and the capacitance of a line where the field is confined in the dielectric layer of a substrate assumed to be of permittivity $\varepsilon_r - 1$.

The air capacitance has already been calculated. In order to calculate the second capacitance mentioned above, the dielectric-air interfaces are replaced by magnetic walls, and the permittivity is set to $\varepsilon_r - 1$. The region corresponding to the dielectric layer in Fig. 3.4a is transformed into the lower x half-plane (Fig. 3.4c) by means of the mapping $x = \cosh^2(\pi z/2h)$ and then into the rectangular domain of Fig. 3.4d via the mapping:

$$w = \int_{x_0}^{x} \frac{dx}{\sqrt{(x-1) \cdot (x-x_1) \cdot (x-x_2) \cdot (x-x_3)}}$$

The layer distributed capacitance is therefore:

$$C_{la}' = 2 \cdot \varepsilon_0 \cdot \frac{d_{12}'}{d_{23}'} = 2 \cdot \varepsilon_0 \cdot (\varepsilon_r - 1) \cdot \frac{K(k_2)}{K(k_2')} \tag{3.2}$$

where:

$$k_2 = \frac{\sinh(\frac{\pi a}{2h})}{\sinh(\frac{\pi b}{2h})} \cdot \sqrt{\frac{1 - \frac{\sinh^2(\frac{\pi b}{2h})}{\sinh^2(\frac{\pi c}{2h})}}{1 - \frac{\sinh^2(\frac{\pi a}{2h})}{\sinh^2(\frac{\pi c}{2h})}}}$$

Now the values for C'_{AIR} and C'_{SUB} when *there is no metal thickness* are:

$$C'_{AIR} = C'_a$$

$$C'_{SUB} = C'_a + C'_{la}$$

If the thickness of the metals is not zero, there is a need to add another component to the distributed air capacitance. This corresponds to the parallel plate capacitors formed by the vertical edges of the center metal and ground metals coupling to each other, and will be included later.

The equivalent circuit for the model of coplanar waveguide in the ISiT technology (Fig. 3.1) is shown in Fig. 3.5.

The capacitance through the substrate (C'_{SUB}) is composed of different portions in ISiT. Imagine that there is a unit positive charge on the center metal and a unit negative charge on the ground metal, and consider an electric field line that originates from the center metal and terminates on the ground metal. This line passes first through the silicon oxide dielectric layer on top, then enters the bulk silicon, and then again enters the silicon oxide layer under the ground metal on one of the two sides, and finally terminates on one of the two return ground metals. Because most of the lines behave in this way, C'_{SUB} can be divided into a series connection of three capacitors. One is due to the silicon oxide under the center metal (C'_{OXSIG} in Fig. 3.1). The other (C'_{SI}) is due to the bulk silicon and has a value equal to the C'_{SUB} calculated above, because for this capacitor the dielectric spatial arrangement is similar to the C'_{SUB} case. The last one is due to the oxide layer directly under the ground metal layer (C'_{OXGND}).

As Fig. 3.5 shows, there is another capacitor which is parallel to the whole above mentioned series connection called C'_N. This capacitor is due to the field lines that do not enter the silicon, but remain the whole way inside the oxide layer. Because

Fig. 3.5 The model for ISiT coplanar *line*

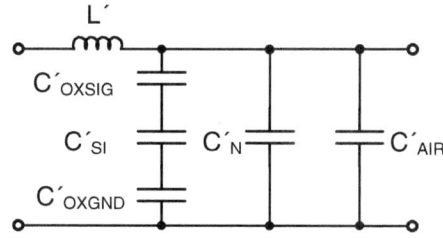

of two reasons it was assumed that the number of these lines is much fewer than the rest (please attention that here number does not mean a physical number, because in reality there are no distinguishable field lines. It means the total electric flux):

1. This layer is very thin (2 μm), compared with the distance between the signal track and the ground plates (at least 10 μm). Therefore, many more lines will enter the bulk silicon instead of staying inside this layer.
2. The electric field lines near the surface of a metal are all perpendicular to the surface in a static situation. Therefore, these field lines don't have enough room to bend towards the ground metal in the thin dielectric layer, so they are forced to enter the bulk silicon.

The fact that the lines are perpendicular to the metal surface has another very important implication: It can be assumed that the lines that are perpendicular to the metals remain still perpendicular (at least approximately) at the interface between the silicon and oxide materials to this interface, as the oxide layer is quite thin. As a result this interface can be approximated as an equipotential surface that behaves like a metal. So it acts like a capacitor electrode, helping defining three distinguishable capacitors with distinguished plates which are in series.

In order to assess the different parameters of the model, the only parameters which remain to be calculated are C'_{OXSIG} and C'_{OXGND}. C'_{AIR} was already calculated earlier by conformal mapping, with some modifications due to metal thickness (will be explained later in this chapter). C'_N is neglected (explained above) and C'_{SI} is the same as the C'_{SUB} in the common model (without the oxide layer), and is calculated also by conformal mapping.

As explained before C'_{OXSIG} is the capacitance between the center metal stripe and the interface between the oxide layer and the silicon.

For calculating C'_{OXSIG}, let's assume the static condition, that is, when there is a positive charge on the center metal and a negative charge on the ground metal. Because the center metal is narrow, the charges will position themselves on its entire surface (on the edges more than in the middle, but this is not important). Therefore, this arrangement can be treated as a parallel plate capacitor. However, due to the metal thickness and the bigger width of the interface between the oxide layer and silicon, which resembles the arrangement of a finite plate on top of an infinitely wide plate, C'_{OXSIG} cannot be calculated directly from the width of the center metal. In order to find out a way to calculate this capacitor by the familiar parallel plate capacitor formula[1], the steps shown in Fig. 3.6 are taken.

As it is shown in the figure, the problem is similar to a thick metal stripe on top of another infinite plane. In our case, this "infinite" plane is simply the interface between the oxide layer and the silicon material, as the interface area here is much wider than center metal width.

In order to consider the effect of the metal thickness, a formula presented in [4] is used. For the new width (w_m), which is the width of a metal stripe with zero thickness

---

[1] $= \varepsilon_0 \varepsilon_r \frac{A}{d}$, ε_0 is the vacuum permittivity (8.85×10^{-12} F/m), ε_r is the dielectric constant, A is the electrode area and d is the electrodes distance.

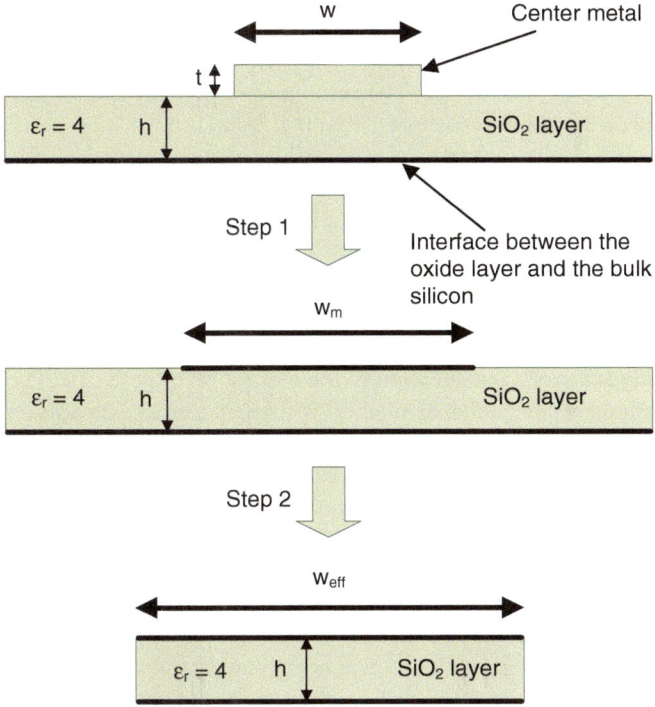

Fig. 3.6 Steps to calculate C'_{OXSIG}

which has the same capacitive coupling to the other plate as the original metal stripe (look at the figure), according to [4]:

$$w_m = w + \frac{1.25 \cdot t}{\pi} \cdot (1 + ln(\frac{4\pi \cdot w}{t}))$$

Now in order to turn the whole arrangement into an equivalent familiar parallel plate structure, with the two plates having the same width:

$$w_{eff} = \frac{2h}{\pi} \cdot \frac{\pi w_m}{2h} + 1 + ln(2\pi \cdot (\frac{w_m}{2h} + 0.92))$$

Finally there is a configuration for which the distributed capacitance can be easily calculated (the last in Fig. 3.6).

For the parameters in the formulas above see Fig. 3.6. The distributed C'_{OXSIG} will be (parameter units are shown in square brackets):

$$C'_{OXSIG} \ [F/m] = \frac{\varepsilon_0 \ [F/m] \cdot \varepsilon_r \cdot w_{eff} \ [\mu m]}{2 \ \mu m}$$

3.1.1 Calculating C'_{OXGND}

Calculation of C'_{OXGND} is different. The assumption that in the static case, the charges are distributed on the entire surface of the ground metal is not always correct. Because when this metal is relatively large and positive charges are put on the center metal and negative charges are put on the ground metal, the negative charges will accumulate around the edge near to the center metal. This accumulation is due to the attractive static forces between the different charges. Consequently, the charge can definitely not cover the entire ground metal surface.

One point that is worth explaining here is why the approach of putting positive charge on the center metal and negative charge on the ground metal is discussed here. Generally for every arrangement of the metals, in order to calculate the capacitance between them, the standard method is to put a positive charge on one of the two metals (does not matter which), and negative charge on the other. The potential difference between the two metals is calculated by integrating the electric field on the path between the separated charges. The capacitance is the inverse of the potential difference per unit charge. This quasi-static capacitor maintains its value approximately up to frequencies around 40 GHz [3].

Let's go back to the discussion about C'_{OXGND}. Definitely this capacitance has a value proportional to the area of the portion of the ground metal covered by the negative charge. The area of the ground metal not covered by charge does not contribute to the capacitance value because of the definition of the capacitance. But how this area portion can be calculated?

Charge and electric field are in direct relation to each other. In order to find out to which extent the negative charge on the ground plane extends, the problem can be regarded like this: To which extent the electric field lines continue to terminate on the ground metal?

Consider first that there is no ground metal, and the center metal is observed from the position of the ground metal in its absence. In this way the center metal will look like a line of charge. As it is known from basic electromagnetics, the electric field of a line of charge is attenuated with a rate equal to the inverse of the perpendicular distance from the line. So as the distance between the electric field lines is inversely proportional to the electric field intensity, the distance between the electric field lines is linearly proportional to the distance from the line. Therefore, in the absence of the ground metal, because the electric field is attenuated proportionally to the inverse of the distance from the center metal, the distance between the field lines increases linearly with the increasing distance from the center metal.

In the presence of the ground metal with negative charge, the situation becomes different. Now the ground metal, on which the field lines terminate, acts as an absorbing medium for them. So they will converge due to the ground metal, the conclusion:

Around the ground metal, the field lines will no longer diverge (get far from each other) with a rate proportional to the distance from center metal, but with a slower rate. That is, the field lines get far from each other not as fast as a linear function with respect to the distance from the center metal, but with a lower order.

So it can be assumed that the distance of the field lines from each other, at the location of the ground plate, will increase like a function that has an order smaller than one with respect to distance from the center metal. C'_{OXGND} is proportional to the width of the portion of the ground metal on which the field lines terminate. This width depends in turn on the distance of the field lines at the location of the ground metal. If the field lines are farther from each other, this width is bigger, and if they are closer to each other, this width is smaller. Therefore, C'_{OXGND} will also increase, with an increasing distance from the center metal, with a rate which is smaller than a linear dependence on the distance. In other words, C'_{OXGND} depends on the slot width (the distance between the center metal and the ground metal, abbreviated s in the following), with a relation in which s has an order lower than one.

At this point, no formula or relationship describing this dependence is available in theory. It was tried to determine the relationship through simulations in the following.

3.2 Simulations for Coplanar Lines

For calculating C'_{OXGND}, a very large amount of EM simulation was performed. The simulation software used was Sonnet.

Simulations were done for center metal widths of 30–90 μm. For every metal width, the slot width (s) was changed from 10 to 350 μm, in order to cover the different necessary values for the characteristic impedance of the line (between 35 and 100 Ω). The ground metal had always a constant width of 500 μm in all the simulations, as its effect was considered to be negligible. Other simulations showed that C'_{OXGND} does not depend on the ground metal width as long as it is larger than 1.5 times the ground to ground distance.

The parametric simulation capability of Sonnet was used.

Although the simulation ranges are limited, for example the center metal width had a maximum of 90 μm in the simulations, the final extracted model predicted the simulation results of structure geometries outside above ranges quite accurately.

For each set of simulations in which the center metal width was kept constant and the slot width was changed from 10 to 350 μm, three different substrates were used in three sets of simulations:

1. In the first set of simulations (defined above), the coplanar waveguide structure was simulated with air (i.e. $\varepsilon_r = 1$) as the surrounding dielectric (everywhere). From the characteristic impedance value of the line with this substrate, which is given by the simulator, the distributed inductance is calculated using the following formula for all parameter combinations.

$$L' = \frac{Z_{AIR}}{c}$$

In the above formula c is the speed of light in air (or vacuum). So as L' will not change with substrate, from this simulation set the distributed inductance of the coplanar lines are extracted in order to be used in the next two sets of simulations. From the resulting L' and Z_{AIR}, C'_{AIR} (see Figs. 3.2, 3.3 and 3.5) was calculated, using the following procedure (s is the slot width):

$$C'_{totalAIR} = \frac{L'}{Z^2_{AIR}}$$

Which is the definition of the characteristic impedance, and then:

$$C'_{AIR} \ [F/m] = \frac{C'_{totalAIR} \ [F/m]}{2} + \frac{2\varepsilon_0 \cdot 1.5 \ \mu m}{s \ [\mu m]}$$

The division by two in the first expression on the right side is because C'_{AIR} is just the capacitance due to the top half-space in the line model. The second term is the capacitance due to the thickness of the metal. The metal thickness capacitance is treated like a parallel plate capacitor. Only half of the metalization thickness (1.5 of 3 μm) is considered due to the fact that the other half has already been included implicitly in the first term ($C'_{totalAIR}$ / 2), but because this capacitance exists on both sides of the center metal, it is also multiplied by 2.

2. In the second step, a homogeneous substrate with a thickness of 508 μm and a dielectric constant of 11.9 was used for the simulation set. From the resulting characteristic impedance and the previously calculated C'_{AIR}, C'_{SUB} of Fig. 3.3 which is equal to C'_{SI} in the third model (Fig. 3.5) was extracted. Now only two unknown variables for the coplanar line on the layered structure (Fig. 3.5) remain: C'_{OXSIG} and C'_{OXGND}.

3. The last substrate for the simulation set was the ISiT substrate, having a 2 μm thick dielectric layer with dielectric constant of 4 on a 508 μm thick material with dielectric constant of 11.9. Here again, using L' (from the first step) and the resulting values from the simulation for the characteristic impedance, the distributed capacitance was calculated. Then with subtracting C'_{AIR} from this value, the value for the series connection of C'_{OXGND}, C'_{SI} and C'_{OXSIG} of Fig. 3.5 was extracted. Then using C'_{SI} values from the second step and the values for C'_{OXSIG} from the theoretical formulas (see above), the values of C'_{OXGND} are computed. Therefore, C'_{OXGND} is derived from the three sets of simulations on three different substrates, each consisting of several subsets, in which every subset has a constant center metal width and varying slot width.

The above calculations were done using MATLAB software. In the implemented MATLAB program, the simulation results for the characteristic impedance from the three substrate sets were the inputs, and the values for C'_{OXGND} were calculated. Each time the program calculates the C'_{OXGND} for a fixed center metal width and a slot width varying from 10 to 350 μm.

Fig. 3.7 C'_{OXGND} values in pF, depending on the slot width, here the center metal width is 70 μm: **a** The fit *curve (solid line)*, **b** results from the EM simulations (x signs)

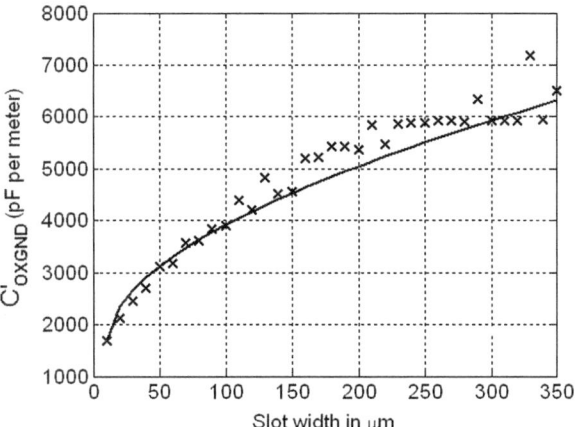

As it is foreseen by theory, C'_{OXGND} increases with the slot width with a rate lower than linear. In Fig. 3.7, the calculated C'_{OXGND} and a fit curve to its values are shown.

So as it is seen in Fig. 3.7, it is possible to find a curve with a simple formula which fits the data. For this case with a center conductor with 70 μm width, the formula for the fit curve is:

$$C'_{OXGND} \, [pF/m] = 1700 \, [pF/m] + 187 \cdot (s \, [\mu m] - 10 \, \mu m)^{0.55}$$

The center metal width is not always 70 μm, and a general formula should be found, which includes the center metal width as a parameter, and gives C'_{OXGND} values for different center metal and slot width values.

For every simulation set in which the center metal width is constant and s is changing, a curve similar to the one in Fig. 3.7 is acquired. The interesting point is that the fit-function has always the same form. Below you see the fit-functions for several center metal widths:

For center metal width of 30 μm:

$$C'_{OXGND} \, [pF/m] = 1200 \, [pF/m] + 111 \cdot (s \, [\mu m] - 10 \, \mu m)^{0.55}$$

For center metal width of 50 μm:

$$C'_{OXGND} \, [pF/m] = 1540 \, [pF/m] + 166 \cdot (s \, [\mu m] - 10 \, \mu m)^{0.55}$$

For center metal width of 90 μm:

$$C'_{OXGND} \, [pF/m] = 1826 \, [pF/m] + 207 \cdot (s \, [\mu m] - 10 \, \mu m)^{0.55}$$

So these fitting-functions have the general formula:

$$A(w) \ [pF/m] + B(w) \cdot (s \ [\mu\text{m}] - 10 \ \mu\text{m})^{0.55}$$

w is the center metal width. In order to find the general formula that evaluates C'_{OXGND} as a function of s and w parameters simultaneously, the parameters A and B are curve-fit with an appropriate function which depends on w.

The fitting functions found for A and B are:

$$A(w) \ [pF/m] = 1200 \ pF/m + 79 \cdot (w \ [\mu\text{m}] - 30 \ \mu\text{m})^{0.5}$$

$$B(w) = 111 + 12 \cdot (w \ [\mu\text{m}] - 30 \ \mu\text{m})^{0.5}$$

So the following general formula holds for C'_{OXGND}:

$$C'_{OXGND} \ [pF/m] = 1200 \ pF/m + 79 \cdot (w \ [\mu\text{m}] - 30 \ \mu\text{m})^{0.5}$$
$$+ (111 + 12 \cdot (w \ [\mu\text{m}] - 30 \ \mu\text{m})^{0.5}) \cdot (s \ [\mu\text{m}] - 10 \ \mu\text{m})^{0.55} \tag{3.3}$$

Now all the parameters needed to calculate the distributed capacitance in the ISiT technology are available. As mentioned before (Sect. 3.1), from the conformal mapping method C'_{AIR} and C'_{SI} are calculated. As thick metals are present here the parallel plate capacitance due to this thickness must be added to the calculated C'_{AIR} from this method. The final C'_{AIR} is equal to:

$$C'_{AIR} \ [F/m] = C'_a \ [F/m] + \frac{2\varepsilon_0 \ [F/m] \cdot 3 \ \mu\text{m}}{s \ [\mu\text{m}]}$$

C'_a was calculated by Eq. 3.1. In the above formula, the number 3 is the thickness of the gold layer in μm.

Please attention that the capacitor C'_N (Fig. 3.5) is neglected because it is very small.

The formula for calculating the distributed inductance is available in the literature. As was explained before, the distributed inductance does not depend on the dielectric material around the coplanar line. Another method for calculating the distributed inductance is mentioned in Sect. 4.1. Having the distributed capacitance (C') and inductance (L'), the effective dielectric constant[2] ($\varepsilon_{r,eff} = \frac{C'}{C'_{totalAIR}}$) and the charac-teristic impedance of the line ($Z_0 = \sqrt{\frac{L'}{C'}}$) can be calculated. Up to now the model is able to deliver these two important parameters of the coplanar lines. In Figs. 3.8, 3.9, 3.10 and 3.11, the comparison between the Sonnet simulation and the developed model results is illustrated.

---

[2] $\varepsilon_{r,eff} = \frac{c}{v}$, where c is the speed of light in vacuum and v is the speed of the signal on the coplanar line with the presence of the substrate.

Fig. 3.8 Characteristic
impedance versus slot width
of **a** developed model (*solid
line*), **b** EM simulations for
the ISiT substrate (+ signs),
c EM simulations for a
homogeneous substrate with
dielectric constant of 11.9
(*circles*), all simulations
correspond to center metal
width of 30 μm

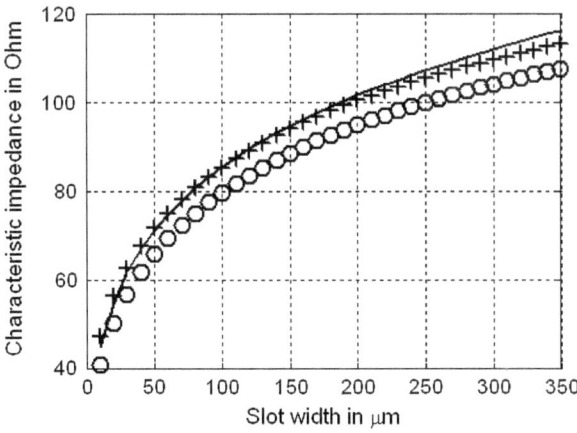

Fig. 3.9 Effective dielectric
constant versus slot width of
a developed model (*solid
line*), **b** EM simulations for
the ISiT substrate (+ signs),
c EM simulations for a
homogeneous substrate with
dielectric constant of 11.9
(*circles*), all simulations
correspond to center metal
width of 30 μm

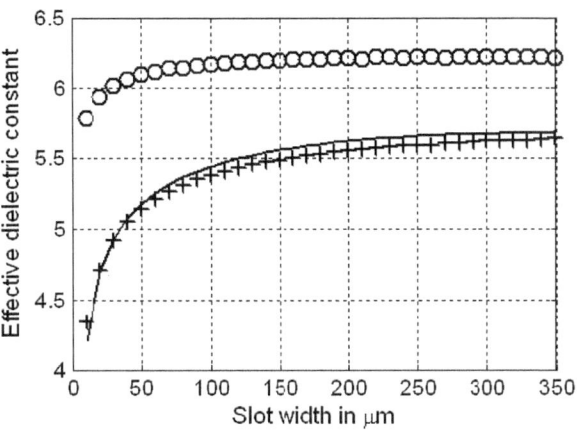

Fig. 3.10 Characteristic
impedance versus slot width
of **a** developed model (*solid
line*), **b** EM simulations for
the ISiT substrate (+ signs),
c EM simulations for a
homogeneous substrate with
dielectric constant of 11.9
(*circles*), all simulations
correspond to center metal
width of 70 μm

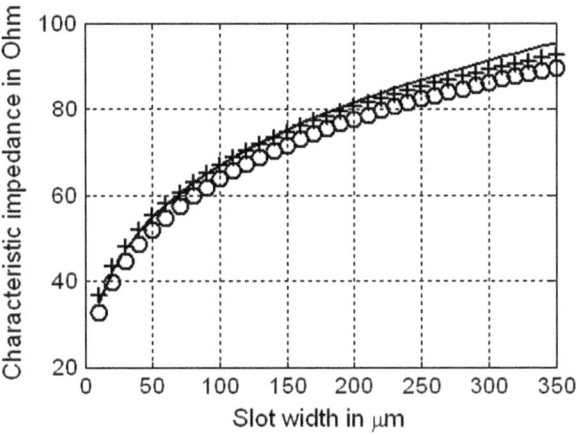

Fig. 3.11 Effective dielectric
constant versus slot width of
a developed model (*solid
line*), **b** EM simulations for
the ISiT substrate (+ signs),
c EM simulations for a
homogeneous substrate with
dielectric constant of 11.9
(*circles*), all simulations
correspond to center metal
width of 70 μm

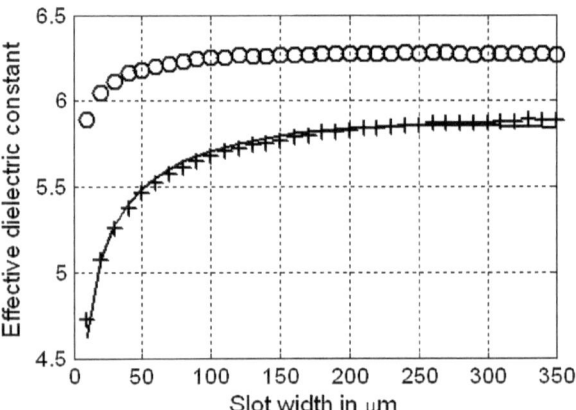

As is seen in these figures, especially for the effective dielectric constant, the
model shows a very good accuracy. The model was checked for geometries out of
the simulation ranges used for extracting the model and it proved to be accurate
there as well. If you look at the curves for the effective dielectric constant, you see
that here there is a big difference between a coplanar line on a homogeneous silicon
substrate and on an ISiT substrate. For example, for a center metal width of 70 μm
and a slot width of 35 μm (a typical structure with a characteristic impedance near
50 Ω), from Fig. 3.11, the difference between the two dielectric constants is about
16 %. So not considering the oxide layer on the substrate would have meant a 16 %
error in estimating the effective dielectric constant.

Calculating the dielectric constant accurately is particularly important for developing RF MEMS phase shifters. There, it is crucial to have an accurate assessment
for the effective dielectric constant.

As it was explained in Chap. 1, the measured loss of the coplanar lines on the ISiT
substrate having poly-silicon is quite near to the one predicted by the ADS coplanar
waveguide model, whose substrate is lossless and determined by conductor ohmic
losses. The attenuation coefficient calculation (i.e. the signal losses) for the coplanar
line in ADS is based on the Wheeler's incremental inductance formula [4]:

$$\alpha_c^{cw} \ [dB/unit \ length] = 4.88 \times 10^{-4} \cdot R_s \cdot \varepsilon_{r,eff} \cdot Z_0 \cdot \frac{P'}{\pi s} \cdot (1 + \frac{w}{s}) \cdot$$
$$\frac{\frac{1.25}{\pi} ln \frac{4\pi w}{t} + 1 + \frac{1.25 t}{\pi w}}{[2 + \frac{w}{s} - \frac{1.25 t}{\pi s} \cdot (1 + ln \frac{4\pi w}{t})]^2}$$

Here R_s is the surface resistivity of the conductors, Z_0 is the characteristic
impedance, $\varepsilon_{r,eff}$ is the effective dielectric constant, t is the metal thickness, w is
the center metal width and s is the slot width.

P' is given by:

$$P' = (\frac{K}{K'})^2 \cdot P$$

Here K is $K(k)$ and K' is $K(k')$ which are both complete elliptic integrals of the first kind with respect to k (ratio of center metal width to the ground to ground distance of the coplanar line) and:

$$k' = \sqrt{1 - k^2}$$

P is defined as follows:

$$P = \frac{k}{(1 - \sqrt{1 - k^2}) \cdot (1 - k^2)^{0.75}} \quad \text{for} \quad 0 \leq k \leq 0.707$$

$$P = \frac{1}{(1 - k) \cdot \sqrt{k}} \cdot (\frac{K'}{K})^2 \quad \text{for} \quad 0.707 \leq k \leq 1$$

As it is seen, these formulas use the characteristic impedance and the effective dielectric constant to calculate the attenuation constant. At this point satisfying formulas for calculating the characteristic impedance and the effective dielectric constant for the coplanar line on the ISiT substrate have been achieved. With the above formulas for losses, the line model is complete and all the necessary parameters for the line are available.

3.3 Measurement Results and Conclusion

In this chapter a new model for the coplanar line on the ISiT substrate was developed. In the literature there is no model for such a structure in which a dielectric layer with a different dielectric constant exists on top of the silicon. As was seen, the values calculated by the new model for the characteristic impedance and effective dielectric constant are in very good agreement with EM simulation results. Furthermore, using the available formulas from the literature and the newly developed formulas for calculating characteristic impedance and effective dielectric constant, a method was proposed for calculating the losses. As mentioned before, the poly-silicon layer suppresses the inversion channel losses completely.

Figure 3.12 shows the measured effective dielectric constant for two fabricated coplanar waveguides with characteristic impedances of 50 and 80 Ω. As it can be seen the developed model can predict the measurements to a reasonable extent. For comparison, the calculated effective dielectric constant by the ADS model by setting the substrate dielectric constant to 10 and 11.9 is illustrated. Setting to 10 is done to approximate the effect of the dielectric constant of the oxide layer. The developed

Fig. 3.12 CPW effective
dielectric permittivity. Above:
$w = 130\,\mu\text{m}, s = 63\,\mu\text{m},$
$Z_0 = 50\,\Omega,$ below:
$w = 40\,\mu\text{m}, s = 108\,\mu\text{m},$
$Z_0 = 80\,\Omega$

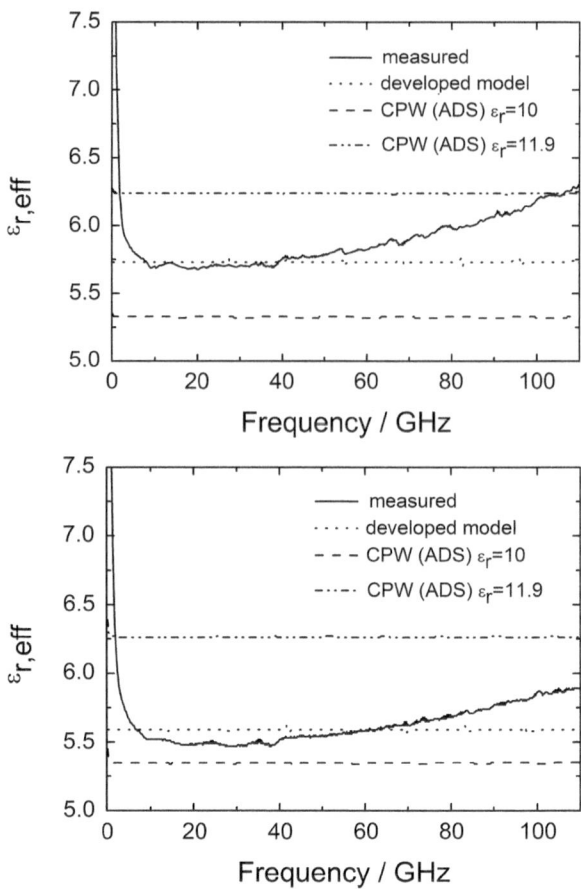

model cannot predict dispersion, which is the dependence of the effective dielectric
constant (alternatively the signal speed) on the frequency. The reason is that here the
quasi-static method was used for calculating the effective dielectric constant, which
is valid only till 40 GHz frequency.

Chapter 4
DC-Block Modeling

DC-block (which is an MIM capacitor) is another component that is used in the circuits. Here in the following, a DC-block is modeled which may have different combinations of dielectric layers (a single AlN layer, a single SiN layer, or a combination of both, refer to Chap. 1 for explanation of the layers), and different geometrical dimensions (overlap length, center metal width, ground metal width, ...). In the literature there are numerous articles that simply extract the model elements (like ideal capacitor, resistor, inductor, ... which build the equivalent circuit of the MIM-capacitor) from the measured S-parameters of the built capacitors. However none of these approaches calculates the model elements directly from the MIM capacitor's geometrical characteristics. So what is done here is quite a new approach. The proposed algorithm, calculates the model elements directly from the geometrical characteristics of the capacitor (via mathematical formulas). This approach is very flexible as the model predicts the behavior of an infinite set of capacitors with different geometrical parameters.

The typical model for an MIM capacitor (DC-block), at least for the frequencies below the resonance frequency, is a model as shown in Fig. 4.1.

As it will be explained later, it was found that this model, although showing good functionality, cannot explain some of the characteristics of the MIM capacitor of ISiT accurately enough. Therefore, this model was modified and another one that perfectly matches the simulation results was suggested. This latter model has been tested through lots of simulations and it is quite successful in predicting the behavior of the DC-block over frequency (Fig. 4.2). The theory behind this second model is explained later in this chapter.

4.1 Modeling Procedure

"Modeling procedure" here means finding formulas for calculating different components of the equivalent circuit (or model) of the MIM capacitor. At this point Fig. 4.1 is considered as the model for the DC-block, but later we will change to Fig. 4.2.

A DC-block structure is shown in Fig. 4.3. In Fig. 4.4 the cross section of the DC-block is shown. As you see in these two pictures, from the left there is a gold layer which overlaps a dielectric layer in some region. The gold layer is first on the

© The Author(s) 2015
N. Pour Aryan, *Design and Modeling of Inductors, Capacitors and Coplanar Waveguides at Tens of GHz Frequencies*, SpringerBriefs in Electrical and Computer Engineering, DOI 10.1007/978-3-319-10187-3_4

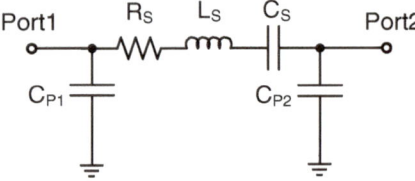

Fig. 4.1 The model for MIM capacitor from [6]

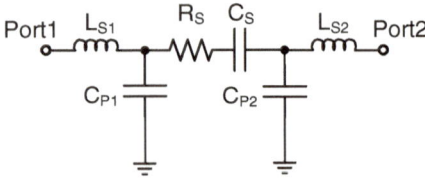

Fig. 4.2 The modified MIM capacitor model suggested here (the inductor has been split into two at the *input* and *output* ports)

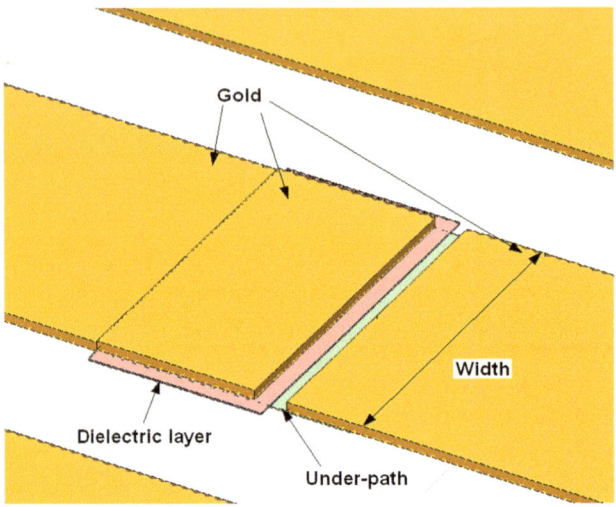

Fig. 4.3 Three dimensional view of DC-block

oxide layer of the substrate and then in the area of the capacitor, it is on a dielectric layer which is itself located on another metal layer. This metal layer, in turn, is on top of the oxide layer of the substrate. This metal layer is made of the UPATH layer of the ISiT technology and extends further to the right, till it touches another gold layer on the right side. The gold layer is on top of the UPATH layer in this area and is physically touching it. In this way, the gold layer on the right connects the right side electrode of the MIM capacitor to the outside world. The capacitor that was included

Fig. 4.4 Cross section view
of DC-block (not to scale)

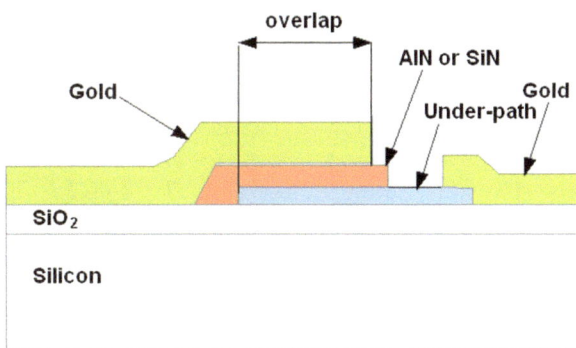

in the design-kit involves the overlapping area, but also two pieces of line to the left
and to the right side, to connect the overlapping area to the outside. This capacitor
is the same as in Fig. 4.3, and the distance between the left edge of the overlapping
area and the left end of the capacitor is 25 μm, whereas the distance from the right
edge of the overlapping area to the right end of the capacitor is 35 μm.

4.1.1 Calculation of C_S

For the modeling procedure let's begin with the main element in the model, i.e.,
the series capacitance. This series capacitance has two components: the parallel
plate capacitance calculated easily from the corresponding formula, and the fringing
capacitance, whose value will be calculated in the immediate following. The fringing
capacitance is due to the electric field lines that exist outside the space between the
two electrodes.

In Table 4.1, for each overlap length and center metal width combination, the EM-
simulated value for the series capacitor, the calculated value for the corresponding
parallel plate capacitor, and the fringing capacitance which is the difference between
the latter two, and the error caused for the capacitor value in percent if this fringing
capacitance is neglected (i.e., the ratio of the fringing capacitance to the capacitance
from the parallel plate structure) are shown. These values are from the simulations
where SiN was used as the dielectric for the MIM capacitor. The overlap length is the
length of the overlapping area between the two metal electrodes (plates), and center
metal width is the width of the signal metal between the two ground metals (which
makes the capacitor). These are shown in Figs. 4.3 and 4.4.

Here in the following, EM-simulations are considered to represent the reality in
a sufficiently accurate manner, which is an approximation. Simulations for many
more possible combinations were done, but due to the space limitation, they are not
included here. As it is seen in the table, the error due to this fringing capacitance
is high for 20 μm overlap, and lower for higher overlaps. The value of the fringing

Table 4.1 Series capacitance parameters for different capacitor geometries with SiN as dielectric

Overlap in μm	Center metal width in μm	C_S from the parallel plate formula in pF	C_S from simulation in pF	Fringing capacitance in pF	Error in neglecting fringing C (%)
20	30	0.1327	0.1375	0.0048	3.58
20	50	0.2213	0.2287	0.0074	3.37
20	70	0.3097	0.32	0.0103	3.31
20	90	0.3982	0.41	0.0118	2.95
20	110	0.4868	0.502	0.0152	3.13
40	30	0.2655	0.27	0.0045	1.69
40	50	0.4425	0.45	0.0075	1.69
40	70	0.6195	0.63	0.0105	1.7
40	90	0.7965	0.81	0.0135	1.69
40	110	0.9735	0.99	0.0165	1.69
60	30	0.3983	0.404	0.0057	1.44
60	50	0.6637	0.673	0.0093	1.39
60	70	0.9293	0.9413	0.0121	1.3
60	90	1.1948	1.2098	0.0151	1.26
60	110	1.4603	1.478	0.0177	1.22

capacitance is also not that much higher for bigger overlaps (when the structures with same center metal width are compared). Therefore, just one model for the fringing capacitance for a structure with an overlap equal to 20 μm was made. It was shown that if this capacitance (which only depends on the center metal width) is added to the capacitance from the parallel plate structure formula, the error (difference from the simulated series capacitance) will always be less than 1 %. Calculations proved that for overlaps up to 200 μm this still holds.

For the other two dielectric combinations, the values for fringing capacitance do not differ that much. If you look at Fig. 4.5, the fringing capacitance values of the structures with one and two dielectric layers are not very different, so one formula (the shown fit line) is used to model this fringing capacitance for all three combinations. The model for the fringing capacitance is simply the equation of this line, or (units of the variables are shown in square brackets):

$$Fringing\ Capacitance\ [pF] = 0.0001125 \cdot Width\ [\mu m] + 0.00225 pF \quad (4.1)$$

This equation holds for center metal widths larger than 30 μm. The fringing capacitance is added to the parallel plate capacitance in the model to give the final value for the series capacitance.

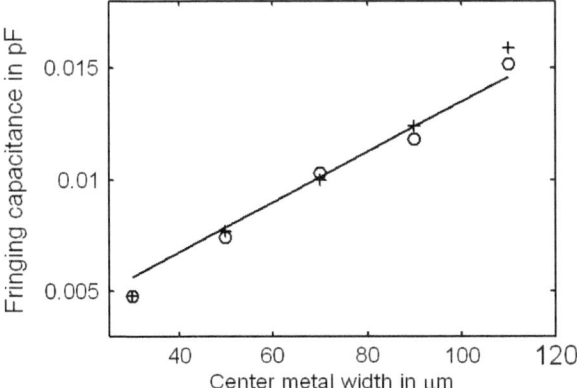

Fig. 4.5 Fringing capacitance values for **a** capacitor with two dielectric layers and overlap of 20 μm (+ signs), **b** capacitor with SiN dielectric layer (one dielectric layer) and overlap of 20 μm (*round circles*), **c** the fit *line* to the simulated data, this will represent the model for the fringing capacitor. As it is seen here the same equation can be used to calculate the fringing capacitance of MIM capacitors with one or two dielectric layers, which is the equation of the *line* shown here

4.1.2 Calculation of C_{P1} and C_{P2}

The elements C_{P1} and C_{P2} in the model account for the capacitances between the capacitor electrodes and the CPW ground planes. In both cases there is a structure in which a metal stripe is between two other metal plates. In the following C_{P1} corresponds to the capacitance due to the left electrode in the structure shown in Fig. 4.3 and C_{P2} corresponds to the right electrode. At the electrode on left, the area on which this electrode is overlapping the other electrode has capacitive coupling to the two ground plates only through the air. This capacitance value can be calculated through a conformal mapping method which was described in Sect. 3.1. In addition, an equivalent parallel-plate capacitance due to the metal thickness must be added (similar to the calculation of C'_{AIR} for coplanar line). As mentioned earlier, the left electrode includes a part of the transmission line to the left, which is directly on the substrate. The length of this part is chosen in the design-kit to be equal to 25 μm. For this part the distributed capacitance is calculated from the line model in the previous chapter. To obtain the final value for C_{P1}, the resulting distributed capacitances were multiplied by the corresponding lengths of the pieces of metal on the left and summed to calculate C_{P1} .

For calculating C_{P2} it should be noted that in the part where the electrode on the right has overlapping to the other electrode, its capacitive coupling to the ground is only through the substrate. The distributed capacitance between the center metal and the ground plates in this case can be calculated by the developed formulas from the previous chapter (this distributed capacitance is equal to the series connection of C'_{SI}, C'_{OXSIG} and C'_{OXGND}, see Fig. 3.5). But also on the right side, there is a piece of coplanar line where the electrode metal does not overlap with the other electrode. This part is treated like a coplanar line, and the distributed capacitance is calculated

Table 4.2 EM-simulated and model-calculated values for C_{P1}

Overlap in μm	Center metal width in μm	C_{P1} from model in fF	C_{P1} from simulation in fF	Ratio of C_{P1} from model to EM simulation
20	30	2.5321	2.51	1.0088
20	50	3.0749	3.01	1.0216
20	70	3.5521	3.47	1.0237
20	90	4.0208	3.9	1.031
20	110	4.5123	4.34	1.0397
60	30	2.8932	3.02	0.958
60	50	3.5024	3.54	0.9894
60	70	4.0419	4.04	1.0005
60	90	4.5749	4.51	1.0144
60	110	5.1377	5.01	1.0255
120	30	3.4350	3.8	0.9039
120	50	4.1435	4.4	0.9417
120	70	4.7765	4.94	0.9669
120	90	5.4061	5.5	0.9829
120	110	6.0758	6.08	0.9993

from the line model of the previous chapter. Finally like C_{P1}, the total capacitance is calculated by multiplying the resulting distributed capacitances by the corresponding lengths of the pieces of metal on the right which sum up to C_{P2}.

To verify the model, EM simulations with Sonnet software were performed. C_{P1} and C_{P2} can be calculated from the simulated Y-parameters via following formulas:

$$C_{P1} = \frac{Imag(Y_{P1})}{\omega} \qquad\qquad C_{P2} = \frac{Imag(Y_{P2})}{\omega} \qquad (4.2)$$

where:

$$Y_{P1} = Y_{11} + Y_{12} \qquad\qquad Y_{P2} = Y_{22} + Y_{21} \qquad (4.3)$$

In Tables 4.2 and 4.3, the simulated and the model-calculated values for C_{P1} and C_{P2} are shown. As it is seen here, the model can be easily implemented in a program to diminish the errors. The values of C_{P1} and C_{P2} were extracted at rather low frequencies because of reasons which will be explained later. The results correspond to a ground to ground distance of 200 μm and a ground metal width of 400 μm.

As you can see in Tables 4.2 and 4.3, the method for calculating C_{P1} and C_{P2} works quite well. The worst error in this set of data for C_{P1} is about 10 % and for C_{P2} is less than 6 %. For other cases the resulting values are quite close to the simulated ones.

Table 4.3 EM-simulated and model-calculated values for C_{P2}

Overlap in μm	Center metal width in μm	C_{P2} from model in fF	C_{P2} from simulation in fF	Ratio of C_{P2} from model to EM simulation
20	30	5.0178	4.76	1.0541
20	50	6.0825	5.8	1.0487
20	70	7.0303	6.72	1.0462
20	90	7.9569	7.6	1.047
20	110	8.9225	8.5	1.0497
60	30	8.4417	8.21	1.028
60	50	10.2393	10	1.0239
60	70	11.8376	11.6	1.0205
60	90	13.3982	13.2	1.015
60	110	15.0222	14.7	1.0219
120	30	13.5777	13.4	1.0133
120	50	16.4745	16.4	1.0045
120	70	19.0484	19.1	0.9973
120	90	21.5601	21.6	0.9982
120	110	24.1717	24.2	0.9988

4.1.3 Series Inductance Calculation

The next parameter to be calculated for the model is the series inductance value. DC-block was considered as a piece of coplanar line with the same signal metal width, ground-to-ground distance and ground metal width. Hence the series inductor was calculated by multiplying the distributed inductance of the corresponding coplanar line by the total length of the capacitor. The distributed inductance is calculated from the coplanar line model in the previous chapter. First, the total distributed capacitance with air as substrate ($C'_{totalAIR}$) is calculated by the method presented in Sect. 3.1, and then the distributed inductance is calculated from the following formula (c is the speed of light in vacuum):

$$L' = \frac{1}{c^2 \cdot C'_{totalAIR}} \tag{4.4}$$

Simulations were done with Sonnet, and the series inductance was calculated from the resulting simulated Y-parameters using the following formulas:

$$L_S = \frac{Imag(\frac{1}{Y_S}) + \frac{1}{\omega \cdot C_S}}{\omega} \tag{4.5}$$

Here C_S is the series capacitance of the model and $Y_S = -Y_{21}$.

Table 4.4 Simulated and model-computed values for the series inductor

Overlap in μm	Center metal width in μm	L_S from model in pH	L_S from simulation in pH	Ratio of L_S from model to EM simulation
20	30	51	49.38	1.0325
20	50	43	41.6	1.0341
20	70	37.6	36.1	1.0402
20	90	33.2	33.7	0.9864
20	110	29.5	28.7	1.0293
60	30	76.5	74.57	1.0256
60	50	64.5	62.87	1.0263
60	70	56.3	54.9	1.026
60	90	49.9	48.8	1.0218
60	110	44.3	43.5	1.0186
120	30	114.7	113	1.0152
120	50	96.8	95.44	1.0141
120	70	84.5	83.4	1.0131
120	90	74.8	74	1.0107
120	110	66.5	65.92	1.0083

It can be proven that the error for the calculated series inductance from the simulation decreases with increasing frequency, so always the largest frequencies immediately below the first resonance frequency were used to extract the inductance from the Y-parameter simulation results.

Finally the simulated and the model-calculated series inductor are compared in Table 4.4.

According to Table 4.4, the worst error for the inductor is about 4 %.

4.1.4 Calculation of R_S

The last parameter to be calculated is the series resistance. It was attempted to calculate this resistance value from the skin depth, but unfortunately the results did not match the EM simulation results.

The reason for this is that here the thickness of the metal is always comparable to the skin-depth (or even smaller), so in fact skin depth concept cannot explain the true distribution of current in the metal, and therefore a solution of Maxwell equations will be necessary. The proximity effect between the center metal and the ground metals affects the losses (see Ref. [1] for details about proximity effect). Currents in the ground plane and in the capacitor electrodes affect each other. These two currents attract each other and therefore the flowing currents tend to accumulate near the edges of the metals. Without using the Maxwell's equations, calculating the resulting current distribution due to this proximity effect is impossible.

Therefore, again curve fitting was used here. It was tried to choose suitable parameters to use for the formulas. Electromagnetic theory is applied again to determine the parameters in a way that the curve-fitting procedure becomes simpler.

A way to model the skin effect in the lumped components is to designate the value of the series resistor in the component's equivalent circuit as follows:

$$R_S = R_{S0} \cdot \sqrt{1 + \frac{f}{f_{g0}}} \qquad (4.6)$$

Here R_{S0} is the DC resistance and f_{g0} is the skin effect corner frequency and is a fitting parameter. When the series resistance is extracted from the EM simulations for different center metal width and overlap combinations of the DC-block, the resulting curves for R_S versus frequency obey the above mentioned formula very well. Therefore, to determine a model for the series resistance it seems that it is enough to extract R_{S0} and f_{g0} values from the simulation curves and find a relationship for them. R_{S0} is extracted from the magnitude of R_S near the frequency of zero. By finding R_{S0}, the other parameter (f_{g0}) is determined from the value of R_S in a relatively high frequency. The series resistance R_S conforms to the above mentioned formula for all possible center metal width and overlap combinations up to 50 GHz.

In order to find a formula for R_{S0}, the process is as follows. From the theoretical point of view, in the first step R_{S0} is obviously linear with respect to the overlap, because the resistance of metals has always a linear relationship with the length of the metal. For the relation with respect to width, a new parameter will be introduced. Because the series resistance should be proportional to the inverse of the area into which the current flows, and the area in which the current flows is proportional to the perimeter of the center metal cross section (because of skin effect), the resistance should be proportional to the inverse of the perimeter. The perimeter itself is equal to:

Perimeter [μm] = 2 · (*Width* [μm] + *Metal thickness* [μm]) = 2 · (*Width* [μm] + 3 μm)

Therefore:

$$R_{S0}\ [\Omega] = \frac{K_1\ [\Omega\mu m]}{Perimeter\ [\mu m]} + cte\ [\Omega] = \frac{K\ [\Omega\mu m]}{Width\ [\mu m] + 3\ \mu m} + cte\ [\Omega]$$

Above *cte* stands for a constant value, $K_1 = 2K$ are values which depend on the overlap. With defining the new variable *IWidth* as follows:

$$IWidth\ [\mu m^{-1}] = \frac{1}{Width\ [\mu m] + 3\ \mu m}$$

Fig. 4.6 R_{S0} versus *IWidth*
for **a** overlap of 20 μm
(*circles*), **b** overlap of 60 μm
(+ signs)

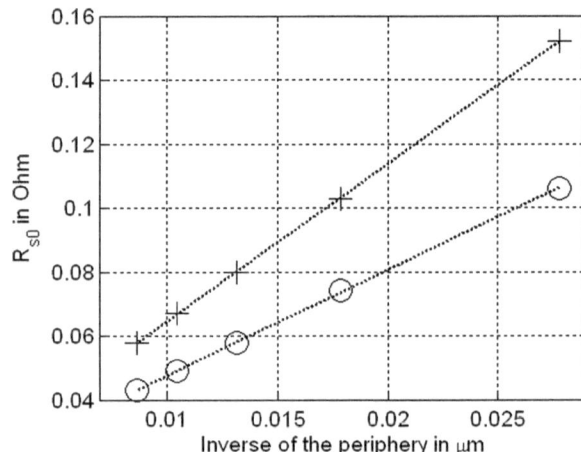

The following relationship between R_{S0} and *IWidth* holds for a fixed overlap:

$$R_{S0}\ [\Omega] = K\ [\Omega\mu m] \cdot IWidth\ [\mu m^{-1}] + cte\ [\Omega]$$

EM simulations confirmed that the curves of R_{S0} versus *IWidth* are linear for a constant overlap, as predicted by the above relationship.

Figure 4.6 shows the simulated R_{S0} versus *IWidth* for two distinct fixed overlaps. Simulations were done for different overlaps. By investigating the results the following relationship between R_{S0} and *IWidth* was established:

$$R_{S0}\ [\Omega] = A\ [\Omega] \cdot overlap\ [\mu m] \cdot IWidth\ [\mu m^{-1}] + 0.0148\ \Omega \qquad (4.7)$$

The value 0.0148 Ω is not a function of the overlap. However, A has a linear dependence on the overlap, which can be described with the following equation:

$$A\ [\Omega] = 0.042\ \Omega\mu m^{-1} \cdot overlap\ [\mu m] + 2.43\ \Omega \qquad (4.8)$$

In the above equation the constant 2.43 Ω corresponds to the two pieces of coplanar line on the two sides of the overlap area, which connect the sandwiched capacitor region to the outside. As stated before these two parts have lengths of 25 and 35 μm.

So finally the following relationship for calculating R_{S0} from center metal width and overlap values is proposed:

$$R_{S0}\ [\Omega] = (0.042\ \Omega \cdot overlap\ [\mu m] + 2.43\ \Omega\mu m) \cdot IWidth\ [\mu m^{-1}] + 0.0148\ \Omega$$
$$(4.9)$$

It was found out that f_{g0} does not depend considerably on the overlap, but it does change with width. Simulations established the following relationship between f_{g0} and *IWidth*:

$$f_{g0} \text{ [GHz]} = 407.5 \text{ GHz}\mu\text{m} \cdot IWidth \text{ } [\mu\text{m}^{-1}] + 6.95 \text{ GHz} \qquad (4.10)$$

Having determined formulas for R_{S0} and f_{g0}, R_S can be calculated for every frequency using Eq. 4.6. The above formulas were used to program the series resistance of the DC-block component of the design-kit in the corresponding model.

At this point, all elements of the model of Fig. 4.1 are calculated. However, this model has a shortcoming:

In the model, in the shunt branches at the input and the output there is only one single capacitor at a time (remember that because in the simulations the substrate is set lossless, no resistor is required here). In the simulations, when the amount of this capacitor was calculated from the Y-parameters (see Eq. 4.2), the resulting values for C_{P1} or C_{P2} depended on frequency. But according to the model, these must be actually independent from the frequency. In Fig. 4.9 the values of C_{P1} over frequency for DC-blocks having different center metal width and overlap combinations are illustrated. The same concept is shown in Fig. 4.10 for C_{P2}.

In Figs. 4.9 and 4.10, at 0 Hz, the values are the same as predicted by the above developed model, but with increasing frequency they change. Their behavior is similar to when there were an inductor in series with them.

The models for MIM capacitors in some literature put the inductor due to the coplanar line nature of the DC-Block in front of the shunt capacitors, i.e. not inside the central branch, with series resistor and capacitor at one place. This is done in order to take into account the distribution effect along the two capacitor electrodes. So here the total inductor was divided into two pieces with approximately equal values. The resulting model is illustrated in Fig. 4.2. This is in agreement with the fact that the C_{P1} and C_{P2} curves versus frequency give the predicted values from the quasi-static calculation of the C_{P1} and C_{P2} at zero frequency, because at zero frequency the inductors are shorted. Consequently, the simulated values for series inductance, series capacitor and resistor (L_S, C_S and R_S) for different DC-blocks agree with the corresponding derived values from the model.

As C_{P1} and C_{P2} are relatively small, the same formulas as before can be used to derive the values of these elements from the EM simulated Y-parameters. Here the formula used for calculating L_S gives now the sum of the two inductors in the input and output of the DC-block. But as is shown later, the frequency dependent behavior of the measured C_{P1} and C_{P2} is predicted very nicely using the new model (Fig. 4.2).

4.2 DC-Block in the Design-Kit

As we will see in the next chapter, a design-kit including different passive elements was created. One of these components is a DC-block with the possibility to be used in S-parameters simulations.

The model that was used for this DC-block is what was developed in this chapter. The above developed formulas were used to program different elements of this model. Its equivalent circuit is the one shown in Fig. 4.2. The model program is available in the Appendix (file name: DC_block_model). In Figs. 4.7, 4.8, 4.9, 4.10 and 4.11, the different parameters extracted from EM simulations are compared to their corresponding values from the schematic simulation of the DC-block component using the developed model. As it is seen the model predicts the behavior of the DC-block satisfactorily.

In Fig. 4.8 the series inductance is the sum of the two inductors on the left and on the right side of the model of Fig. 4.2.

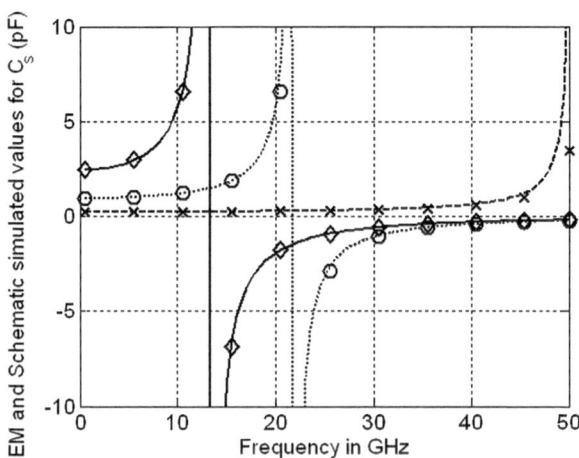

Fig. 4.7 Simulated values for series capacitance of the DC-block for **a** overlap of 20 μm and center metal width of 50 μm from EM simulation (x signs) versus design-kit's DC-block simulation (*dashed line*), **b** overlap of 60 μm and center metal width of 70 μm from EM simulation (*circles*) versus design-kit's DC-block simulation (*dotted line*), **c** overlap of 100 μm and center metal width of 110 μm from EM simulation (*diamonds*) versus design-kit's DC-block simulation (*solid line*)

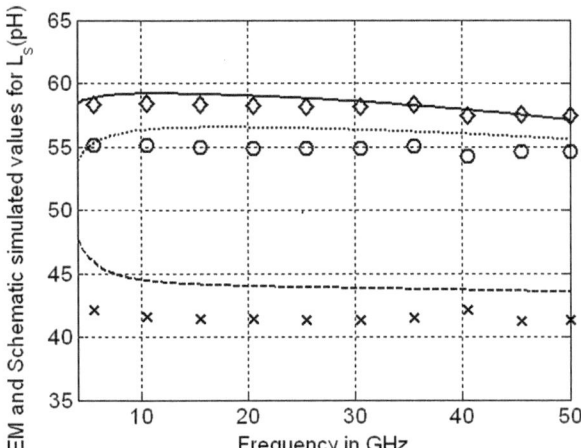

Fig. 4.8 Simulated values for series inductance of the DC-block for **a** overlap of 20 μm and center metal width of 50 μm from EM simulation (x signs) versus design-kit's DC-block simulation (*dashed line*), **b** overlap of 60 μm and center metal width of 70 μm from EM simulation (*circles*) versus design-kit's DC-block simulation (*dotted line*), **c** overlap of 100 μm and center metal width of 110 μm from EM simulation (*diamonds*) versus design-kit's DC-block simulation (*solid line*)

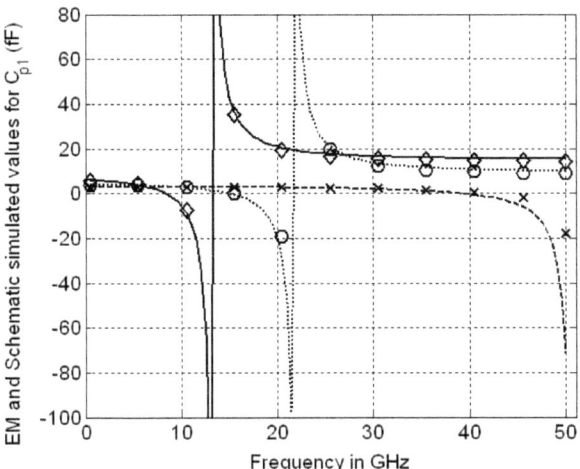

Fig. 4.9 Simulated values for C_{P1} of the DC-block for **a** overlap of 20 μm and center metal width of 50 μm from EM simulation (x signs) versus design-kit's DC-block simulation (*dashed line*), **b** overlap of 60 μm and center metal width of 70 μm from EM simulation (*circles*) versus design-kit's DC-block simulation (*dotted line*), **c** overlap of 100 μm and center metal width of 110 μm from EM simulation (*diamonds*) versus design-kit's DC-block simulation (*solid line*)

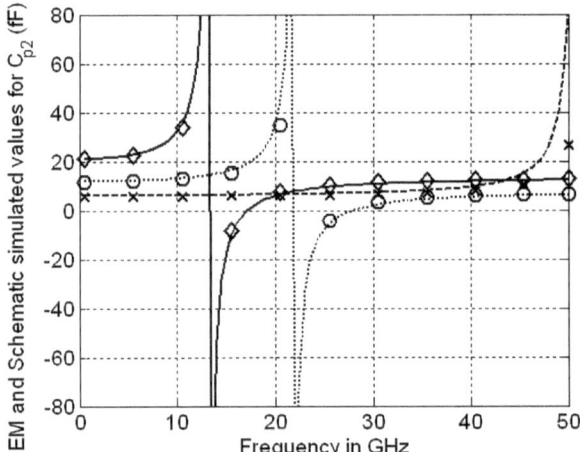

Fig. 4.10 Simulated values for C_{P2} of the DC-block for **a** overlap of 20 μm and center metal width of 50 μm from EM simulation (x signs) versus design-kit's DC-block simulation (*dashed line*), **b** overlap of 60 μm and center metal width of 70 μm from EM simulation (*circles*) versus design-kit's DC-block simulation (*dotted line*), **c** overlap of 100 μm and center metal width of 110 μm from EM simulation (*diamonds*) versus design-kit's DC-block simulation (*solid line*)

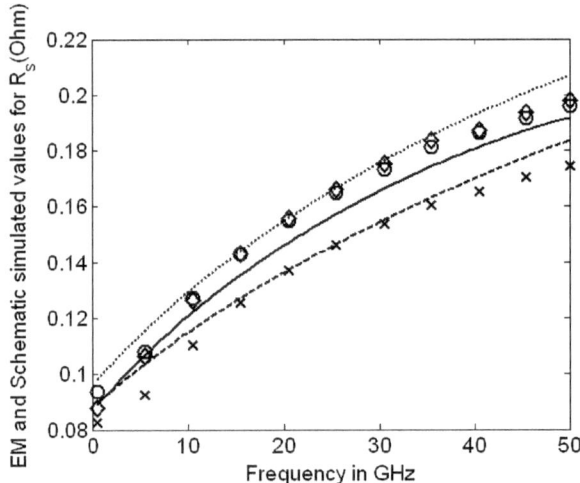

Fig. 4.11 Simulated values for R_S of the DC-block for **a** overlap of 20 μm and center metal width of 50 μm from EM simulation (x signs) versus design-kit's DC-block simulation (*dashed line*), **b** overlap of 60 μm and center metal width of 70 μm from EM simulation (*circles*) versus design-kit's DC-block simulation (*dotted line*), **c** overlap of 100 μm and center metal width of 110 μm from EM simulation (*diamonds*) versus design-kit's DC-block simulation (*solid line*)

As it is shown in Fig. 4.9, the simulated C_{P1} depends on frequency. This is not because this capacitor is really changing with frequency, but because of the presence of an inductor between the input of the DC-block and this capacitor (Fig. 4.2). The formula for calculating C_{P1} from the Y-parameters (Eq. 4.2) results in this behavior due to the presence of this inductor. The same discussion holds for Cp2 (Fig. 4.10).

As it is shown in Fig. 4.11, the model produced by curve fitting is also able to predict the series resistance of the DC-block satisfactorily.

Chapter 5
Design-Kit Programming

In the last chapters different passive elements were modeled or optimized. The best way to use these models and structures is to put them inside a design-kit in ADS. This helps the designers to work fast and efficiently with them.

A design-kit was programmed for ADS. It was called ISiT-kit. The built-in programming language AEL inside ADS was used. AEL stands for: Application Extension Language. AEL is used to configure, customize and extend the capabilities of the design environment. AEL is modeled after the popular programming language C.

Doing this the components discussed above were added to ADS as new palettes (buttons). They appear on the schematic as other ADS components. They all have an associated layout with them that is generated either by using the *Generate/Update layout...* command from the *layout* menu in the schematic window, or simply by going to the layout window and pressing the correspondent component palette from there.

The design-kit has the following components:

- Four ISiT MEMS switches designed by Fraunhofer institute.
- Nine inductors designed in Chap. 2 and grouped into two separate palette lists (for 24 and 35 GHz).
- Scalable modules, including the DC-block (MIM capacitor) modeled in the previous chapter, and two different coplanar lines: one made of LINES layer (gold) and the other made of UPATH layer.

The components are included in 4 palette lists with the names:

1. ISiT switches
2. 24 GHz Optimized ISiT Inductors
3. 35 GHz Optimized ISiT Inductors
4. ISiT Technology Scalable modules

Each palette list has the components' buttons plus a palette for the netlist include component (refer to the ADS documentation for details).

Most of the components, i.e. the inductors, DC-block, and the ISiT switches can be used for S-parameters simulation. The inductors and ISiT switches have the corresponding S-parameter files (in Dataset format for the inductors and Touchstone

© The Author(s) 2015   51
N. Pour Aryan, *Design and Modeling of Inductors, Capacitors and Coplanar Waveguides at Tens of GHz Frequencies*, SpringerBriefs in Electrical and Computer Engineering, DOI 10.1007/978-3-319-10187-3_5

format for the ISiT switches), and the DC-block has a model provided by a Spice program which calculates the different model components (resistance, capacitances,...) from the DC-block's geometrical parameters.

The program files generated are included in the Appendix.

5.1 The Building Blocks of the Design-Kit

The details of building the design-kit are not explained here. Interested reader can refer to the chapters "Design Kit Development", "AEL", ...provided by the ADS documentation.

The first step for a beginner is to go through the section "ADS Design Kit Tutorial" in the book "Design Kit Development" and read it through. In the following some materials are explained that are not included in this tutorial and/or it is very difficult to find them in the ADS documentation. Some tricks are also described.

The design-kit from the tutorial has a mother folder with some name (ISiT_design_kit here). Beside that there are sub-folders which contain AEL files, model files (with .net ending), bitmap files for the palettes, symbol files and others. A list of the sub-folders is given below:

design_kit

doc

circuit/symbols

circuit/ael

circuit/models

circuit/bitmaps/pc

circuit/bitmaps/unix

circuit/records

de/ael

examples

The purpose of these folders and the files inside them are explained in the tutorial. Furthermore, ISiT-kit has also the following additional folders:

5.1.1 Circuit/Artwork

This folder contains either the layout design files (.dsn) of the components or the AEL files which generate layouts for scalable modules, e.g. DC-block and coplanar lines.

When there is a fixed artwork for a component, e.g. the component has a fixed layout like the inductors or the switches designed here, the corresponding layout (.dsn) file is saved inside this folder. Different alternatives for the fixed artwork can be chosen in the create_item() command of the component.

In the case of a scalable layout, an AEL program is written for layout generation. The necessary commands are available in the AEL book in ADS documentation. There are various commands to change the current layer (layout layer), to draw rectangles and ports, etc.

The component should always have ports, otherwise there will be problems when generating layouts from a schematic that has several instances of the same component. The port is used by ADS to locate the component properly in the layout window.

Another way to generate AEL layout generation programs is to use the "Graphical Cell Compiler" in ADS. Please refer to the corresponding ADS documentation for explanation.

5.1.2 Circuit/Data

When some components have S-parameter files (like Dataset) for simulation, this simulation data is added to the design-kit as follows:

A model is defined for the component that will be accessed with the help of the netlist include component. In this model file, a S2P component is instantiated. For its file name, the name of the corresponding S-parameter file is chosen without including any addressing information. The S-parameter file (Touchstone, Dataset, …) is put in the data folder inside the circuit folder of the design-kit. The advantage of this configuration is that there is no addressing information and therefore the design-kit could be transported from one computer to another without worrying about what happens to the addresses.

The method is as follows (which is easier than writing the program for the model):

A S2P component is put inside the schematic and the lower pin (pin 3, the common terminal) is grounded. Then a netlist of this is created by simulating the circuit. The simulator gives an error and exits but the resulting netlist file is now available in the File Browser window. It is opened and the line in which the S2P component was instantiated is copied to the proper place in the model program. The FILE name is changed to the name of the Dataset file of the corresponding component.

5.1.3 de/Defaults

Because the modules contain artworks (the corresponding layouts), the default files for the preferences (.prf) and for the layers (.lay) must be included inside the defaults sub-folder of the de folder of the design-kit. These files are used to define the preferences and the layout layers for the layouts of the components.

5.2 Conclusion

All components have corresponding layouts and all can be used for S-parameter simulation.

The DC-block has a scripted model program associated with it. The model was developed in the previous chapter. The program receives the geometrical parameters from the component (overlap, width of the center metal, distance between the two ground metals, ...) and calculates the values for the elements of the equivalent circuit of the DC-block. The way that the program makes the netlist for the model is based on the PSpice netlisting format and is explained here:

If you look in the Appendix, to where the model file for the DC-block is printed (File name: DC_block_model), you see the following lines, before each **end model_name** line.

L:LSL n1 4 L=LVALUELEFT

C:CS 4 5 C=CSVALUE

Y_Port:Y2P1 5 0 6 0 Y[1,1]=1/RVALUE Y[1,2]=-1/RVALUE Y[2,1]=-1/RVALUE Y[2,2]=1/RVALUE Recip=no

L:LSR 6 n2 L=LVALUERIGHT

C:CP1 4 0 C=CP1VALUE fF

C:CP2 6 0 C=CP2VALUE fF

In the PSpice netlisting format n1 and n2 are the left and the right ports, respectively. The first line is interpreted like this: from the left port to the node number 4, put an inductor with a value equal to the variable LVALUELEFT (which has been previously calculated in the program).

The numbers between 4 and 6 in the above lines designate the nodes (each number one corresponding node). Of course another set of numbers (for example: 3, 5, 7) can be used. The number 0 always represents the ground node.

The programmed model for the DC-block can predict the EM simulations quite accurately (as was shown in Sect. 4.2).

In the next pages, Figs. 5.1, 5.2, 5.3, 5.4, 5.5, 5.6 and 5.7 show different parts of the design-kit through screen-shots.

In Fig. 5.1 one of the inductor components in the schematic window is seen. Here in the circuit two 50 Ω ports are connected to the component for simulation. The component ISIT INCLUDE connects the inductor to its S-parameter simulation data. On the left side of the window 7 palettes are seen. One is for the ISIT INCLUDE component and the others are for the 6 inductors of the 24 GHz operation frequency group.

In Fig. 5.2 some inductors are visible in the layout window. On the left the palettes of the 35 GHz operation frequency group are seen.

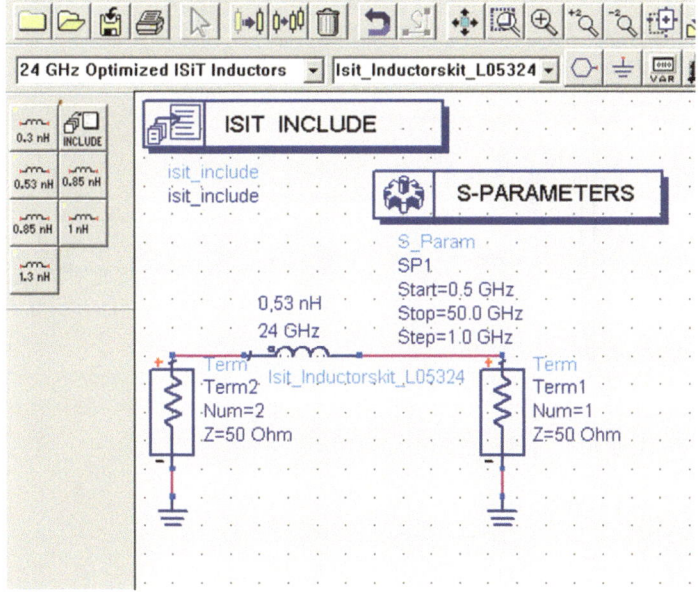

Fig. 5.1   An inductor component in the schematic window

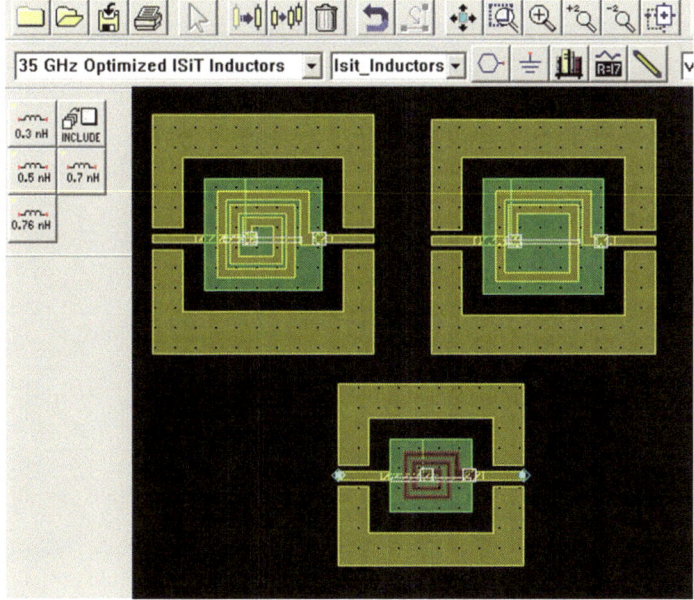

Fig. 5.2   Some design-kit inductors in the layout window

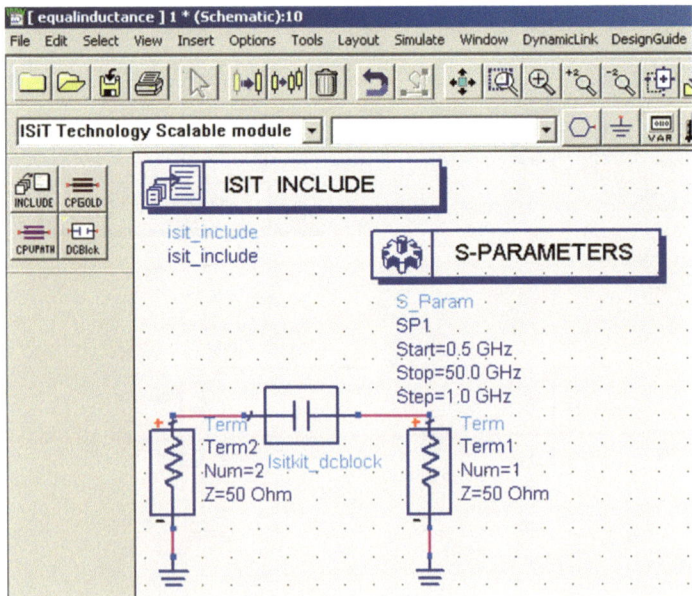

Fig. 5.3 DC-block component in the schematic window

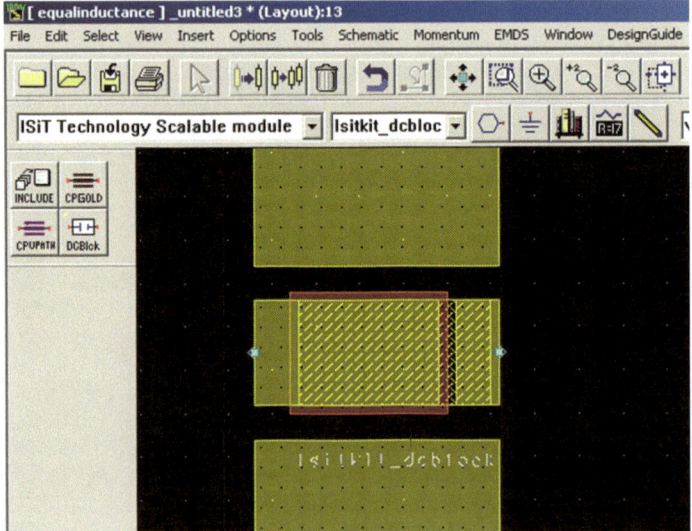

Fig. 5.4 DC-block component in the layout window

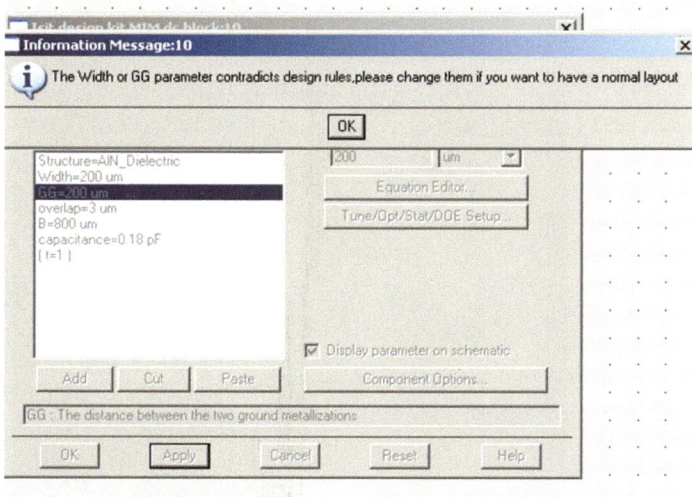

Fig. 5.5 The interactive dialog box of the DC-block component

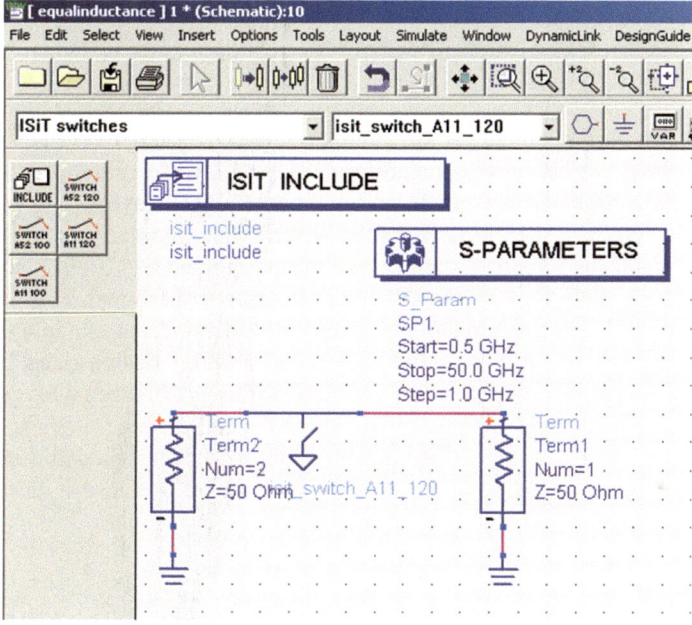

Fig. 5.6 A switch component in the schematic window

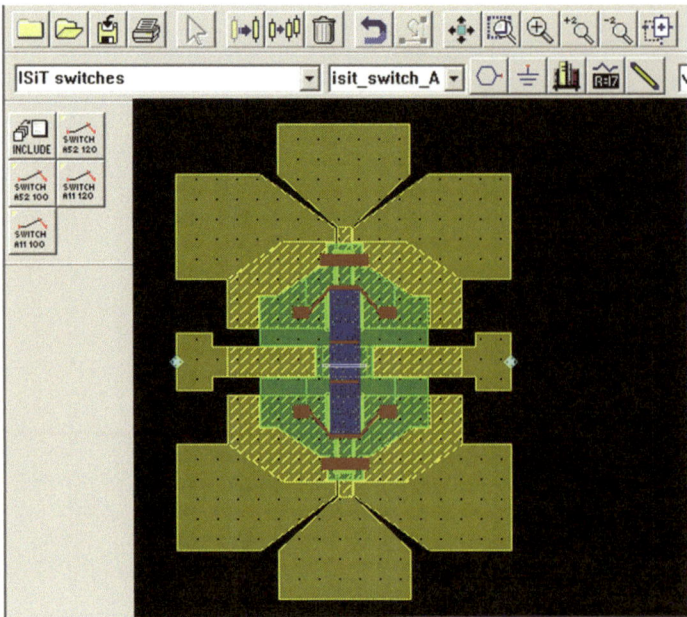

Fig. 5.7 The switch of Fig. 5.6 in the layout window

In Fig. 5.3 an arrangement for simulating a DC-block component is shown. On the left 4 palettes are observable, one for the ISIT INCLUDE, one for the DC-block and two for two coplanar lines made from gold and UPATH layers.

In Fig. 5.4 you see the layout for a DC-block. The geometry of the layout is automatically adjusted by the parameters of the component.

In Fig. 5.5 the dialog box of the DC-block component is shown. This dialog box is interactive. If the input parameters for the DC-block are not suitable for layout generation, for example if they do not obey the technology design-rules, or if there is doubt that the simulation results are accurate, different messages will appear. The messages describe what is not appropriate about the parameters.

In Figs. 5.6 and 5.7, a switch component is shown in the schematic and layout windows. On the left there are for palettes for the switches, because at this time only 4 switches are available in the design-kit.

Chapter 6
Summary

In this essay first it was explained that how a technology introduced by Fraunhofer institute suppresses the losses caused by the inversion channels under the oxide layer of the ISiT substrate. It was shown that this makes it possible to predict the losses of passive elements in this technology by setting the substrate loss to zero in the EM simulations. This in turn assures that the simulated results shown here for the losses are close to reality.

In the following steps two libraries of inductors were designed for the frequencies of 24 and 35 GHz and new models were developed for the coplanar line structure on the ISiT substrate and for the DC-block. These are the passive elements used with the RF MEMS switches in the circuits.

The figures of merit in the inductor design were high quality factor and low area consumption. The coplanar-line model predicts the effective dielectric constant and the characteristic impedance in the ISiT technology quite accurately. The losses could be derived from the available formulas in the literature. The DC-block model is a novel approach, and is developed using electromagnetic concepts along with the help of EM simulations. This model gives the S-parameter behavior of a DC-block having an arbitrary combination of geometrical parameters very precisely.

Later a design-kit including the designed passive components plus several RF MEMS switches was programmed and added to ADS. The components are available in ADS for layout generation or simulation. For layout generation for DC-block an AEL program is written, which receives the different geometrical parameters from the component and sketches the appropriate layout. Inductors and switches though use fixed layout patterns which are stored in the design-kit. The DC-block component uses a scripted model program for simulation, while the inductors and switches use S-parameter Dataset and Touchstone files, respectively.

© The Author(s) 2015

N. Pour Aryan, *Design and Modeling of Inductors, Capacitors and Coplanar Waveguides at Tens of GHz Frequencies*, SpringerBriefs in Electrical and Computer Engineering, DOI 10.1007/978-3-319-10187-3_6

Appendix
Design-Kit Codes

Some of the main files in the design-kit, written with the programming language AEL:

File name: boot

File position: ISiT_design_kit/de/ael

The boot file of the design-kit (the sentences in front of // mean comment):

```
1   // boot.ael - This file resides in the de/ael directory of the
2   // design kit. It is loaded by the design kit infrastructure
3   // software if it is listed in the file ads.lib in one of 4
4   // predefined locations, one of which is $HOME/hpeesof/design_kit.
5
6   // This file is used to load other AEL files such as palette.ael.
7   // It is also used to set up some global variables for use in
8   // other files. One global variable that is available by default:
9   // designKitRecord - this is a list which contains the 4 fields
10  // from ads.lib (kit name, path, boot file, version).
11  // As soon as the design kit load process has finished, this
12  // variable is unset, so save the values as a variable with a
13  // different name if you want access to them later.
14  //
15  // The following debug print statements can be used to view the
16  // values of these variables:
17  //
18  // To print a field in the list:
19  //fputs(stderr, designKitRecord[0]);
20  //de_info(identify_value(designKitRecord[0]));
21  // Now save the path variable so it is available for later use.
22  // Any variables declared in this file must be unique to this
23  // design kit.
24  // If you copy this boot file to make another design kit, these
25  // names must be changed.
26  decl Isitkit_PATH = designKitRecord[1];
27  // To view the path variable:
28  //fputs(stderr, Isit_Inductorskit_PATH);
29  // Comment out all debug print statements before shipping your
30  // design kit.
31  // These path names will be used later to load other files.
32  decl Isitkit_BITMAP_DIR = sprintf("%s/circuit/bitmaps/%s/",
        Isitkit_PATH,
33                          on_PC?"pc":"unix");
```

© The Author(s) 2015
N. Pour Aryan, *Design and Modeling of Inductors, Capacitors and Coplanar
Waveguides at Tens of GHz Frequencies*, SpringerBriefs in Electrical
and Computer Engineering, DOI 10.1007/978-3-319-10187-3

```
34  decl Isitkit_CIRCUIT_AEL_DIR = sprintf("%s/circuit/ael/",
         Isitkit_PATH);
35  decl Isitkit_CIRCUIT_MODEL_DIR = sprintf("%s/circuit/models/",
36  Isitkit_PATH);
37  decl Isitkit_DE_AEL_DIR = sprintf("%s/de/ael/", Isitkit_PATH);
38  // To print a variable:
39  //fputs(stderr,Isit_Inductorskit_BITMAP_DIR);
40  load(strcat(Isitkit_DE_AEL_DIR,"palette"), "CmdOp");
41
42  //the following is for the added layout generator for the
         capacitor
43  //decl Isitkit_1_Root = designKitRecord[1];
44
45  decl Isitkit_1_artwork = sprintf("%s/circuit/artwork/",
         Isitkit_PATH);
46  load(strcat(Isitkit_1_artwork,"
         art_Isitkit_dcblock_modified_art_final"),
47  "CmdOp");
48
49  /////////////////////// Hilfsfunktionen ///////////////////////
50  load(strcat(Isitkit_DE_AEL_DIR,"uudk_isit_generic"), "CmdOp");
51  load(strcat(Isitkit_1_artwork,"isit_artwork"), "CmdOp");
52
53  /////////////////////// Laden der Elemente ///////////////////////
54  load(strcat(Isitkit_CIRCUIT_AEL_DIR,"Isit_Inductorskit_item"),
55  "CmdOp"); load(strcat(Isitkit_CIRCUIT_AEL_DIR,"isit_include"),
56  "CmdOp"); load(strcat(Isitkit_CIRCUIT_AEL_DIR,"isit_cpw_gold"),
57  "CmdOp"); load(strcat(Isitkit_CIRCUIT_AEL_DIR,"isit_cpw_upath"),
58  "CmdOp"); load(strcat(Isitkit_CIRCUIT_AEL_DIR,"isit_switch"),
59  "CmdOp");
60
61  // To print a variable:
62  // fputs(stderr, isit_BITMAP_DIR);
63  decl isit_KITNAME = designKitRecord[0];
```

File name: palette

File Position: ISiT_design_kit/de/ael

The following is the palette code file of the design-kit, producing the palettes for the components in the layout and schematic windows:

```
1   //This file produces the palettes of the design-kit
2   //in schematic and layout windows
3
4   dk_define_palette_group(SCHEM_WIN, "analogRF_net",
5           "24 GHz Optimized ISiT Inductors","24 GHz Optimized
               ISiT Inductors", 0,
6           "Isit_Inductorskit_L0324", "ISIT_INDUCTORSKIT
               Optimized 24GHz Inductor",
7          strcat(Isitkit_BITMAP_DIR,"Isit_Inductorskit_L0324"),
8           "isit_include", "isit process include component",
9              strcat(Isitkit_BITMAP_DIR,"isit_include"),
10                "Isit_Inductorskit_L05324", "
                      ISIT_INDUCTORSKIT Optimized 0.53nH 24GHz
                      Inductor",
11                strcat(Isitkit_BITMAP_DIR,"
                      Isit_Inductorskit_L05324"),
12                "Isit_Inductorskit_L08524", "
                      ISIT_INDUCTORSKIT Optimized 0.85nH 24GHz
                      Inductor",
```

```
13                          strcat(Isitkit_BITMAP_DIR,"
                                Isit_Inductorskit_L08524"),
14                          "Isit_Inductorskit_L08524", "
                                ISIT_INDUCTORSKIT Optimized 0.85nH 24GHz
                                Inductor",
15                          strcat(Isitkit_BITMAP_DIR,"
                                Isit_Inductorskit_L08524"),
16                  "Isit_Inductorskit_L124", "ISIT_INDUCTORSKIT
                        Optimized 1nH 24GHz Inductor",
17                          strcat(Isitkit_BITMAP_DIR,"
                                Isit_Inductorskit_L124"),
18                  "Isit_Inductorskit_L1324", "ISIT_INDUCTORSKIT
                        Optimized 1.3nH 24GHz Inductor",
19                          strcat(Isitkit_BITMAP_DIR,"
                                Isit_Inductorskit_L1324") );
20  dk_define_palette_group(SCHEM_WIN, "analogRF_net",
21              "35 GHz Optimized ISiT Inductors","35 GHz Optimized
                    ISiT Inductors", 1,
22              "Isit_Inductorskit_L0335", "ISIT_INDUCTORSKIT
                    Optimized 0.3nH 35GHz Inductor",
23                  strcat(Isitkit_BITMAP_DIR,"
                        Isit_Inductorskit_L0335"),
24          "isit_include", "isit process include component",
25              strcat(Isitkit_BITMAP_DIR,"isit_include"),
26          "Isit_Inductorskit_L0535", "ISIT_INDUCTORSKIT
                Optimized 0.5nH 35GHz Inductor",
27              strcat(Isitkit_BITMAP_DIR,"
                    Isit_Inductorskit_L0535"),
28          "Isit_Inductorskit_L0735", "ISIT_INDUCTORSKIT
                Optimized 0.7nH 35GHz Inductor",
29              strcat(Isitkit_BITMAP_DIR,"
                    Isit_Inductorskit_L0735"),
30          "Isit_Inductorskit_L07635", "ISIT_INDUCTORSKIT
                Optimized 0.76nH 35GHz Inductor",
31              strcat(Isitkit_BITMAP_DIR,"
                    Isit_Inductorskit_L07635"));
32
33  dk_define_palette_group(SCHEM_WIN, "analogRF_net",
34              "ISiT Technology Scalable modules","ISiT Technology
                    Scalable modules", 2,
35              "isit_include", "isit process include component",
36                  strcat(Isitkit_BITMAP_DIR,"isit_include"),
37              "isit_cpw_gold", "isit gold cpw",
38                  strcat(Isitkit_BITMAP_DIR,"isit_cpw_gold"),
39              "isit_cpw_upath", "isit upath cpw",
40                  strcat(Isitkit_BITMAP_DIR,"isit_cpw_upath"),
41
42                  "Isitkit_dcblock", "Isit kit MIM dc block",
43                      strcat(Isitkit_BITMAP_DIR,"Isitkit_dcblock"))
                            ;
44
45  dk_define_palette_group(SCHEM_WIN, "analogRF_net",
46              "ISiT switches","ISiT switch Components", 0,
47              "isit_include", "isit process include component",
48                  strcat(Isitkit_BITMAP_DIR,"isit_include"),
49              "isit_switch_A52_120", "isit_switch_A52_120",
50                  strcat(Isitkit_BITMAP_DIR,"isit_switch_A52_120"),
51              "isit_switch_A52_100", "isit_switch_A52_100",
52                  strcat(Isitkit_BITMAP_DIR,"isit_switch_A52_100"),
```

```
53              "isit_switch_A11_120", "isit_switch_A11_120",
54                  strcat(Isitkit_BITMAP_DIR,"isit_switch_A11_120"),
55              "isit_switch_A11_100", "isit_switch_A11_100",
56                  strcat(Isitkit_BITMAP_DIR,"isit_switch_A11_100")
57
58  );
59
60  dk_define_palette_group(LAYOUT_WIN, "analogRF_net",
61              "24 GHz Optimized ISiT Inductors","24 GHz Optimized
                ISiT Inductors", 0,
62              "Isit_Inductorskit_L0324", "ISIT_INDUCTORSKIT
                Optimized 24GHz Inductor",
63              strcat(Isitkit_BITMAP_DIR,"Isit_Inductorskit_L0324"),
64              "isit_include", "isit process include component",
65                  strcat(Isitkit_BITMAP_DIR,"isit_include"),
66                  "Isit_Inductorskit_L05324", "
                        ISIT_INDUCTORSKIT Optimized 0.53nH 24GHz
                        Inductor",
67                  strcat(Isitkit_BITMAP_DIR,"
                        Isit_Inductorskit_L05324"),
68                  "Isit_Inductorskit_L08524", "
                        ISIT_INDUCTORSKIT Optimized 0.85nH 24GHz
                        Inductor",
69                  strcat(Isitkit_BITMAP_DIR,"
                        Isit_Inductorskit_L08524"),
70                  "Isit_Inductorskit_L08524", "
                        ISIT_INDUCTORSKIT Optimized 0.85nH 24GHz
                        Inductor",
71                  strcat(Isitkit_BITMAP_DIR,"
                        Isit_Inductorskit_L08524"),
72              "Isit_Inductorskit_L124", "ISIT_INDUCTORSKIT
                Optimized 1nH 24GHz Inductor",
73                  strcat(Isitkit_BITMAP_DIR,"
                        Isit_Inductorskit_L124"),
74              "Isit_Inductorskit_L1324", "ISIT_INDUCTORSKIT
                Optimized 1.3nH 24GHz Inductor",
75                  strcat(Isitkit_BITMAP_DIR,"
                        Isit_Inductorskit_L1324") );
76  dk_define_palette_group(LAYOUT_WIN, "analogRF_net",
77              "35 GHz Optimized ISiT Inductors","35 GHz Optimized
                ISiT Inductors", 1,
78              "Isit_Inductorskit_L0335", "ISIT_INDUCTORSKIT
                Optimized 0.3nH 35GHz Inductor",
79                  strcat(Isitkit_BITMAP_DIR,"
                        Isit_Inductorskit_L0335"),
80              "isit_include", "isit process include component",
81                  strcat(Isitkit_BITMAP_DIR,"isit_include"),
82              "Isit_Inductorskit_L0535", "ISIT_INDUCTORSKIT
                Optimized 0.5nH 35GHz Inductor",
83                  strcat(Isitkit_BITMAP_DIR,"
                        Isit_Inductorskit_L0535"),
84              "Isit_Inductorskit_L0735", "ISIT_INDUCTORSKIT
                Optimized 0.7nH 35GHz Inductor",
85                  strcat(Isitkit_BITMAP_DIR,"
                        Isit_Inductorskit_L0735"),
86              "Isit_Inductorskit_L07635", "ISIT_INDUCTORSKIT
                Optimized 0.76nH 35GHz Inductor",
87                  strcat(Isitkit_BITMAP_DIR,"
                        Isit_Inductorskit_L07635"));
```

```
88
89   dk_define_palette_group(LAYOUT_WIN, "analogRF_net",
90               "ISiT Technology Scalable modules","ISiT Technology
                    Scalable modules", 2,
91               "isit_include", "isit process include component",
92                   strcat(Isitkit_BITMAP_DIR,"isit_include"),
93               "isit_cpw_gold", "isit gold cpw",
94                   strcat(Isitkit_BITMAP_DIR,"isit_cpw_gold"),
95               "isit_cpw_upath", "isit upath cpw",
96                   strcat(Isitkit_BITMAP_DIR,"isit_cpw_upath"),
97
98                   "Isitkit_dcblock", "Isit kit MIM dc block",
99                       strcat(Isitkit_BITMAP_DIR,"Isitkit_dcblock"))
                            ;
100
101  dk_define_palette_group(LAYOUT_WIN, "analogRF_net",
102              "ISiT switches","ISiT switch Components", 0,
103              "isit_include", "isit process include component",
104                  strcat(Isitkit_BITMAP_DIR,"isit_include"),
105              "isit_switch_A52_120", "isit_switch_A52_120",
106                  strcat(Isitkit_BITMAP_DIR,"isit_switch_A52_120"),
107              "isit_switch_A52_100", "isit_switch_A52_100",
108                  strcat(Isitkit_BITMAP_DIR,"isit_switch_A52_100"),
109              "isit_switch_A11_120", "isit_switch_A11_120",
110                  strcat(Isitkit_BITMAP_DIR,"isit_switch_A11_120"),
111              "isit_switch_A11_100", "isit_switch_A11_100",
112                  strcat(Isitkit_BITMAP_DIR,"isit_switch_A11_100")
113  );
```

File name: isit_cpw_gold
File Position: ISiT_design_kit/circuit/ael
The following code produces the LINES coplanar line component for the design-kit:

```
1   //This file produces the LINES coplanar line component for the
       design-kit
2
3   set_simulator_type(1); create_item("isit_cpw_gold",
4    "isit_cpw coplanar gold",
5    "CPW",NULL,NULL,NULL,
6    standard_dialog,"",
7    CmpModelNetlistFmt, "",
8    ComponentAnnotFmt,
9    "SYM_isit_cpw_gold",
10   macro_artwork,"isit_cpw_gold_artwork",
11   ITEM_PRIMITIVE_EX,
12   create_parm("Model", "Model instance name",
13        PARM_NOT_EDITED | PARM_NO_DISPLAY,
14        "StdFileFormset", UNITLESS_UNIT, prm("StdForm","cpw_gold")
            ),
15
16   create_parm("length","the length of the conduction layer",
17            PARM_OPTIMIZABLE | PARM_STATISTICAL,
18            "StdFileFormSet",LENGTH_UNIT,prm("StdForm","100um"),
19            list(dm_create_cb(PARM_MODIFIED_CB,"
                   uudk_isit_check_range_cb","length isit_cpw_gold",
                   TRUE))),
20   create_parm("width","the width of the conduction layer",
21            PARM_OPTIMIZABLE | PARM_STATISTICAL,
22            "StdFileFormSet",LENGTH_UNIT,prm("StdForm","40um "),
```

```
23        list(dm_create_cb(PARM_MODIFIED_CB,"
              uudk_isit_check_range_cb","width isit_cpw_gold",TRUE
              ))),
24    create_parm("spacing","spacing betwwenconduction layer and groud
         ",
25            PARM_OPTIMIZABLE | PARM_STATISTICAL,
26            "StdFileFormSet",LENGTH_UNIT,prm("StdForm","20um"),
27            list(dm_create_cb(PARM_MODIFIED_CB,"
              uudk_isit_check_range_cb","spacing isit_cpw_gold",
              TRUE))),
28    create_parm("ground_width","width of ground layer",
29            PARM_OPTIMIZABLE | PARM_STATISTICAL,
30             "StdFileFormSet",LENGTH_UNIT,prm("StdForm","20um"),
31            list(dm_create_cb(PARM_MODIFIED_CB,"
              uudk_isit_check_range_cb","ground_width
              isit_cpw_gold",TRUE)))
32  );
```

File name: isit_cpw_upath

File Position: ISiT_design_kit/circuit/ael

The following code produces the coplanar line component from UPATH layer for the design-kit:

```
1  //This file produces the coplanar line component from UPATH layer
      for the design-kit
2
3  set_simulator_type(1); create_item("isit_cpw_upath",
4    "isit_cpw coplanar upath",
5    "CPW",NULL,NULL,NULL,
6    standard_dialog,"",
7    CmpModelNetlistFmt, "",
8    ComponentAnnotFmt,
9    "SYM_isit_cpw_upath",
10   macro_artwork,"isit_cpw_upath_artwork",
11   ITEM_PRIMITIVE_EX,
12   create_parm("Model", "Model instance name",
13         PARM_NOT_EDITED | PARM_NO_DISPLAY,
14          "StdFileFormset", UNITLESS_UNIT, prm("StdForm","cpw_upath"
              )),
15    create_parm("length","the length of the conduction layer",
16            PARM_OPTIMIZABLE | PARM_STATISTICAL,
17            "StdFileFormSet",LENGTH_UNIT,prm("StdForm","100um"),
18            list(dm_create_cb(PARM_MODIFIED_CB,"
              uudk_isit_check_range_cb","length isit_cpw_upath",
              TRUE))),
19   create_parm("width","the width of the conduction layer",
20            PARM_OPTIMIZABLE | PARM_STATISTICAL,
21            "StdFileFormSet",LENGTH_UNIT,prm("StdForm","40um "),
22            list(dm_create_cb(PARM_MODIFIED_CB,"
              uudk_isit_check_range_cb","width isit_cpw_upath",
              TRUE))),
23   create_parm("spacing","spacing betwwenconduction layer and groud
         ",
24            PARM_OPTIMIZABLE | PARM_STATISTICAL,
25            "StdFileFormSet",LENGTH_UNIT,prm("StdForm","20um"),
26            list(dm_create_cb(PARM_MODIFIED_CB,"
              uudk_isit_check_range_cb","spacing isit_cpw_upath",
              TRUE))),
27   create_parm("ground_width","width of ground layer",
```

```
28        PARM_OPTIMIZABLE | PARM_STATISTICAL,
29         "StdFileFormSet",LENGTH_UNIT,prm("StdForm","20um"),
30        list(dm_create_cb(PARM_MODIFIED_CB,"
             uudk_isit_check_range_cb","ground_width
             isit_cpw_upath",TRUE)))
31 );
```

File name: isit_include

File Position: ISiT_design_kit/circuit/ael

The following code produces the netlist include component for the design-kit:

```
1  //This file produces the netlist include component for the design-
      kit
2
3  set_simulator_type(1); defun isit_include_netlist_cb (cbP,
4  clientData, callData) {
5      decl fileName="", netlistString="";
6      fileName = strcat(Isit_Inductorskit_CIRCUIT_MODEL_DIR, "
           isit_models.net");
7    netlistString=strcat(netlistString, "#ifndef ISIT_INCLUDE\n");
8    netlistString=strcat(netlistString, "#define ISIT_INCLUDE\n");
9
10     netlistString=strcat(netlistString, "#include \"", fileName,"
           \"\n");
11     fileName="";
12     fileName = strcat(Isit_Inductorskit_CIRCUIT_MODEL_DIR, "
           Isit_Inductorskit_models.net");
13     netlistString=strcat(netlistString, "#include \"", fileName,"
           \"\n");
14     fileName="";
15     fileName = strcat(Isit_Inductorskit_CIRCUIT_MODEL_DIR, "
           dc_block_model.net");
16     netlistString=strcat(netlistString, "#include \"", fileName,"
           \"\n");
17
18     netlistString=strcat(netlistString, "#endif\n");
19     return(netlistString);
20 } create_item("isit_include",                        // name
21          "isit process include components",      // label
22          "isit_include",                          // prefix
23          ITEM_UNIQUE|ITEM_NOT_NETLIST_IF_SUB,    // attribute
24          0,                                       // priority
25          NULL,                                    // iconName
26          standard_dialog,                         // dialogName
27          NULL,                                    // dialogData
28          ComponentNetlistFmt,                     //
                netlistFormat
29          "isit_include",                          // netlistData
30          ComponentAnnotFmt,                       //
                displayFormat
31          "SYM_isit_include",                      // symbolName
32          no_artwork,                              // artworkType
33          NULL,                                    // artworkData
34          0,                                       // extraAttrib
35          list (dm_create_cb (ITEM_NETLIST_CB, "
                isit_include_netlist_cb", NULL, TRUE)));
```

File name: isit_switch

File Position: ISiT_design_kit/circuit/ael

The following code produces the switch components for the design-kit:

```
 1  //This file produces the switch components for the design-kit
 2
 3  //First switch
 4  set_simulator_type(1);
 5  //create_constant_form(name, label, attribute, netlistFormat,
        displayFormat);
 6  create_constant_form("s1", "off state of the switch", 0,
 7  "isit_switch_A52_120_OFF", "Switch off"); create_constant_form("s2
        ",
 8  "on state of the switch", 0, "isit_switch_A52_120_ON", "switch on"
        );
 9  create_form_set("isit_switch_A52_120_formset", "s1","s2");
10  create_item("isit_switch_A52_120","isit_switch","L",ITEM_NORMAL
11  ,-1,"",standard_dialog,"", CmpModelNetlistFmt,   ""
12  ,"","SYM_isit_switch",1,"isit_switch_A52_120.dsn",
        ITEM_CKT_MODEL_EX,
13  create_parm("Model","Model instance
14  name",262144,"isit_switch_A52_120_formset",-1,prm("StdForm","
        isit_switch_A52_120_ON")));
15
16  //Second switch
17  create_constant_form("t1", "off state of the switch", 0,
18  "isit_switch_A52_100_OFF", "Switch off"); create_constant_form("t2
        ",
19  "on state of the switch", 0, "isit_switch_A52_100_ON", "switch on"
        );
20  create_form_set("isit_switch_A52_100_formset", "t1","t2");
21  create_item("isit_switch_A52_100","isit_switch","L",ITEM_NORMAL
22  ,-1,"",standard_dialog,"", CmpModelNetlistFmt,   ""
23  ,"","SYM_isit_switch",1,"isit_switch_A52_100.dsn",
        ITEM_CKT_MODEL_EX,
24  create_parm("Model","Model instance
25  name",262144,"isit_switch_A52_100_formset",-1,prm("StdForm","
        isit_switch_A52_100_ON")));
26
27  //Third switch
28  create_constant_form("u1", "off state of the switch", 0,
29  "isit_switch_A11_120_OFF", "Switch off"); create_constant_form("u2
        ",
30  "on state of the switch", 0, "isit_switch_A11_120_ON", "switch on"
        );
31  create_form_set("isit_switch_A11_120_formset", "u1","u2");
32  create_item("isit_switch_A11_120","isit_switch","L",ITEM_NORMAL
33  ,-1,"",standard_dialog,"", CmpModelNetlistFmt,   ""
34  ,"","SYM_isit_switch",1,"isit_switch_A11_120.dsn",
        ITEM_CKT_MODEL_EX,
35  create_parm("Model","Model instance
36  name",262144,"isit_switch_A11_120_formset",-1,prm("StdForm","
        isit_switch_A11_120_ON")));
37
38  //Fourth switch
39  create_constant_form("v1", "off state of the switch", 0,
40  "isit_switch_A11_100_OFF", "Switch off"); create_constant_form("v2
        ",
41  "on state of the switch", 0, "isit_switch_A11_100_ON", "switch on"
        );
42  create_form_set("isit_switch_A11_100_formset", "v1","v2");
43
```

```
44  create_item("isit_switch_A11_100","isit_switch","L",ITEM_NORMAL
45  ,-1,"",standard_dialog,"", CmpModelNetlistFmt,    ""
46  ,"","SYM_isit_switch",1,"isit_switch_A11_100.dsn",
        ITEM_CKT_MODEL_EX,
47  create_parm("Model","Model instance
48  name",262144,"isit_switch_A11_100_formset",-1,prm("StdForm","
        isit_switch_A11_100_ON")));
```

File name: Isit_Inductorskit_item
File Position: ISiT_design_kit/circuit/ael
The following code produces the inductor components and the DC-block for the
design-kit. It also includes the program which runs the interactive behavior of the
DC-block component parameters' dialog box:

```
1   //This file produces the inductor components for the design-kit
2   //and also the DC-block component, it also includes the program
3   //which runs the interactive behavior of the component's
        parameters'
4   //dialog box.
5
6   set_simulator_type(1);
7   create_item("Isit_Inductorskit_L0324","Isit_Inductorskit_L
        Inductor
8   ,L=0.3 nH, Q at 24 GHz :10","L",ITEM_NORMAL
9   ,-1,"",standard_dialog,"",CmpModelNetlistFmt,"","","
        SYM_Isit_Inductorskit_L0324",1,"art_Isit_Inductors_L0324.dsn",
        ITEM_CKT_MODEL_EX,
10  create_parm("Model","Model instance
11  name",262144,"StdFormSet",-1,prm("StdForm","Isit_Inductors_L0324")
        ));
12  create_item("Isit_Inductorskit_L05324","Isit_Inductorskit_L
13  Inductor ,L=0.53 nH, Q at 24 GHz :11","L",ITEM_NORMAL
14  ,-1,"",standard_dialog,"",CmpModelNetlistFmt,"","","
        SYM_Isit_Inductorskit_L05324",1,"art_Isit_Inductors_L05324.dsn
        ",ITEM_CKT_MODEL_EX,
15  create_parm("Model","Model instance
16  name",262144,"StdFormSet",-1,prm("StdForm","Isit_Inductors_L05324"
        ));
17  create_item("Isit_Inductorskit_L08524","Isit_Inductorskit_L
18  Inductor ,L=0.85 nH, Q at 24 GHz :11","L",ITEM_NORMAL
19  ,-1,"",standard_dialog,"",CmpModelNetlistFmt,"","","
        SYM_Isit_Inductorskit_L08524",1,"art_Isit_Inductors_L08524.dsn
        ",ITEM_CKT_MODEL_EX,
20  create_parm("Model","Model instance
21  name",262144,"StdFormSet",-1,prm("StdForm","Isit_Inductors_L08524"
        ));
22  create_item("Isit_Inductorskit_L124","Isit_Inductorskit_L
        Inductor
23  ,L=1 nH, Q at 24 GHz :9","L",ITEM_NORMAL
24  ,-1,"",standard_dialog,"",CmpModelNetlistFmt,"","","
        SYM_Isit_Inductorskit_L124",1,"art_Isit_Inductors_L124.dsn",
        ITEM_CKT_MODEL_EX,
25  create_parm("Model","Model instance
26  name",262144,"StdFormSet",-1,prm("StdForm","Isit_Inductors_L124"))
        );
27  create_item("Isit_Inductorskit_L1324","Isit_Inductorskit_L
        Inductor
28  ,L=1.3 nH, Q at 24 GHz :4","L",ITEM_NORMAL
```

```
29  ,-1,"",standard_dialog,"",CmpModelNetlistFmt,"","","
        SYM_Isit_Inductorskit_L1324",1,"art_Isit_Inductors_L1324.dsn",
        ITEM_CKT_MODEL_EX,
30  create_parm("Model","Model instance
31  name",262144,"StdFormSet",-1,prm("StdForm","Isit_Inductors_L1324")
        )));
32  create_item("Isit_Inductorskit_L0735","Isit_Inductorskit_L
        Inductor
33  ,L=0.7 nH, Q at 35 GHz :4","L",ITEM_NORMAL
34  ,-1,"",standard_dialog,"",CmpModelNetlistFmt,"","","
        SYM_Isit_Inductorskit_L0735",1,"art_Isit_Inductors_L0735.dsn",
        ITEM_CKT_MODEL_EX,
35  create_parm("Model","Model instance
36  name",262144,"StdFormSet",-1,prm("StdForm","Isit_Inductors_L0735")
        )));
37  create_item("Isit_Inductorskit_L07635","Isit_Inductorskit_L
38  Inductor ,L=0.76 nH, Q at 35 GHz :3","L",ITEM_NORMAL
39  ,-1,"",standard_dialog,"",CmpModelNetlistFmt,"","","
        SYM_Isit_Inductorskit_L07635",1,"art_Isit_Inductors_L07635.dsn
        ",ITEM_CKT_MODEL_EX,
40  create_parm("Model","Model instance
41  name",262144,"StdFormSet",-1,prm("StdForm","Isit_Inductors_L07635"
        )));
42  create_item("Isit_Inductorskit_L0535","Isit_Inductorskit_L
        Inductor
43  ,L=0.5 nH, Q at 35 GHz :8","L",ITEM_NORMAL
44  ,-1,"",standard_dialog,"",CmpModelNetlistFmt,"","","
        SYM_Isit_Inductorskit_L0535",1,"art_Isit_Inductors_L0535.dsn",
        ITEM_CKT_MODEL_EX,
45  create_parm("Model","Model instance
46  name",262144,"StdFormSet",-1,prm("StdForm","Isit_Inductors_L0535")
        )));
47  create_item("Isit_Inductorskit_L0335","Isit_Inductorskit_L
        Inductor
48  ,L=0.3 nH, Q at 35 GHz :9","L",ITEM_NORMAL
49  ,-1,"",standard_dialog,"",CmpModelNetlistFmt,"","","
        SYM_Isit_Inductorskit_L0335",1,"art_Isit_Inductors_L0324.dsn",
        ITEM_CKT_MODEL_EX,
50  create_parm("Model","Model instance
51  name",262144,"StdFormSet",-1,prm("StdForm","Isit_Inductors_L0335")
        )));
52
53  //From this part of the program on,it is just about the DC block
54  set_simulator_type(1); create_constant_form("cf1", "This structure
55  consists of a 300nm thick layer of AlN as dielectric", 0,
56  "ALN_STRUCTURE_SECTION", "AlN_Dielectric");
57  create_constant_form("cf2", "This structure consists of a 300nm
58  thick layer of SiN as dielectric", 0, "SiN_STRUCTURE_SECTION",
59  "SiN_Dielectric"); create_constant_form("cf3", "This structure
60  consists of a 300nm thick layer of AlN plus a 300nm thick layer of
61  SiN on top as dielectric", 0, "TWOLAYER_STRUCTURE_SECTION", "AlN+
        SiN
62  Dielectric"); create_form_set("dcblockFormset", "cf1","cf2","cf3")
        ;
63  create_item("Isitkit_dcblock",
64              "Isit design kit MIM dc block",
65              "DCB",ITEM_NORMAL, -1, "",
66              standard_dialog, "",
67              CmpModelNetlistFmt , "",
```

```
68              "",
69              "SYM_Isitkit_dcblock",
70              macro_artwork, "pam_art_Isitkit_dcblock_modifiedn",
71              ITEM_CKT_MODEL_EX,
72          create_parm("Structure", "Structure instance name",
                PARM_STRING ,
73              "dcblockFormset",UNITLESS_UNIT,prm("StdForm","
                AlN_Dielectric"),list(dm_create_cb( PARM_MODIFIED_CB
                ,"dcblockfunction","Structure",TRUE))),
74          create_parm("Width","The width of the DC block",
75              PARM_OPTIMIZABLE | PARM_STATISTICAL ,
76              "StdFileFormSet",5,prm("StdForm","30 um"), list(
                    dm_create_cb( PARM_MODIFIED_CB,"dcblockfunction","
                    Width",TRUE))),
77          create_parm("GG",
78              "The distance between the two ground metallizations",
79              PARM_OPTIMIZABLE | PARM_STATISTICAL , "StdFileFormSet"
                    ,5,prm("StdForm","100 um"),list(dm_create_cb(
                    PARM_MODIFIED_CB,"dcblockfunction","GG",TRUE))),
80          create_parm("overlap", "The overlap length between the two
                plates of the MIM capacitor",
81              PARM_OPTIMIZABLE | PARM_STATISTICAL ,
82              "StdFileFormSet",5,prm("StdForm","19 um"),list(
                    dm_create_cb( PARM_MODIFIED_CB,"dcblockfunction","
                    overlap",TRUE))),
83          create_parm("B",
84              "The distance between the two outer edges of the ground
                    metallizations",
85              PARM_OPTIMIZABLE | PARM_STATISTICAL , "StdFileFormSet"
                    ,5,prm("StdForm","400 um"),list(dm_create_cb(
                    PARM_MODIFIED_CB,"dcblockfunction","B",TRUE))),
86          create_parm("capacitance", "The capacitance of the component
                ",
87              PARM_OPTIMIZABLE | PARM_STATISTICAL ,
88              "StdFileFormSet",CAPACITANCE_UNIT,prm("StdForm","0.177 pF
                    "),list(dm_create_cb( PARM_MODIFIED_CB,"
                    dcblockfunction","capacitance",TRUE))),
89          create_parm("t", "This parameter controls the layout
                generator",
90              PARM_NO_DISPLAY | PARM_NOT_ON_SCREEN_EDITABLE |
                    PARM_NOT_EDITED | PARM_INT ,
91              "StdFileFormSet",UNITLESS_UNIT,prm("StdForm","1"),list(
                    dm_create_cb( PARM_MODIFIED_CB,"dcblockfunction","t"
                    ,TRUE)))
92      ) ;
93
94  //The following function runs the interactive behavior of the DC-
        block's dialog box
95  defun dcblockfunction(cbP, clientData, callData) { decl
96  dependentParmData = NULL; decl Width = pcb_get_mks(callData,
97  "Width"); decl overlap = pcb_get_mks(callData, "overlap"); decl GG
        =
98  pcb_get_mks(callData, "GG"); decl B = pcb_get_mks(callData, "B");
99  decl capacitance = pcb_get_mks(callData, "capacitance"); decl mod
        =
100 pcb_get_form_value(callData, "Structure");
101 if((strcmp(clientData,"Width") == 0) && (Width < 30 um))
102 de_info("The model for the simulation will not be valid for widths
```

```
103  less than 30 um", 0); if((strcmp(clientData,"Width") == 0) && (
         Width
104  > 100 um)) de_info("The model for the simulation will not be valid
105  for widths more than 100 um", 0); if(((strcmp(clientData,"B") ==
         0)
106  || (strcmp(clientData,"GG") == 0)) && ( B<4*GG )) de_info("The
         model
107  for the simulation will not be valid for B values below 4 times
         the
108  ground to ground distance", 0); if(((strcmp(clientData,"GG") == 0)
109  || (strcmp(clientData,"Width") == 0)) && ( Width > GG-20um))
110  de_info("The Width or GG parameter contradicts design rules,please
111  change them if you want to have a normal layout", 0);
112
113
114
115  if ( ( strcmp(mod, "This structure consists of a 300nm thick layer
116  of AlN as dielectric") == 0) && ((strcmp(clientData,"Width") == 0)
117  || (strcmp(clientData,"Structure") == 0))) {
118      dependentParmData = pcb_set_mks(dependentParmData, "overlap",
             1e-6*round(1e6*((capacitance-(112.5*Width +0.00225)*1e-12)
             * 0.3e-6)/(Width*8.85e-11)));
119      overlap=1e-6*round(1e6*((capacitance-(112.5*Width +0.00225)*1e
             -12)* 0.3e-6)/(Width*8.85e-11));
120  } else if (( strcmp(mod, "This structure consists of a 300nm thick
121  layer of AlN as dielectric") == 0) && (strcmp(clientData,"overlap"
         )
122  == 0))
123
124      dependentParmData = pcb_set_mks(dependentParmData, "
             capacitance", 1e-14*round(1e14*(((overlap*Width*8.85e-11)
             /0.3e-6)+(112.5*Width +0.00225)*1e-12)));
125  else if (( strcmp(mod, "This structure consists of a 300nm thick
126  layer of AlN as dielectric") == 0) &&
127  (strcmp(clientData,"capacitance") == 0)) {
128      dependentParmData = pcb_set_mks(dependentParmData, "overlap",
             1e-6*round(1e6*((capacitance-(112.5*Width +0.00225)*1e-12)
             * 0.3e-6)/(Width*8.85e-11)));
129      overlap=1e-6*round(1e6*((capacitance-(112.5*Width +0.00225)*1e
             -12)* 0.3e-6)/(Width*8.85e-11));
130  } else if (( strcmp(mod, "This structure consists of a 300nm thick
131  layer of SiN as dielectric") == 0) && ((strcmp(clientData,"Width")
132  == 0)|| (strcmp(clientData,"Structure") == 0))) {
133      dependentParmData = pcb_set_mks(dependentParmData, "overlap", 1
             e-6*round(1e6*((capacitance-(112.5*Width +0.00225)*1e-12)*
             0.3e-6)/(Width*7.5*8.85e-12)));
134      overlap=1e-6*round(1e6*((capacitance-(112.5*Width +0.00225)*1e
             -12)* 0.3e-6)/(Width*7.5*8.85e-12));
135  } else if (( strcmp(mod, "This structure consists of a 300nm thick
136  layer of SiN as dielectric") == 0) && (strcmp(clientData,"overlap"
         )
137  == 0))
138      dependentParmData = pcb_set_mks(dependentParmData, "
             capacitance", 1e-14*round(1e14*(((overlap*Width*7.5*8.85e
             -12)/0.3e-6)+(112.5*Width +0.00225)*1e-12)));
139
140  else if (( strcmp(mod, "This structure consists of a 300nm thick
141  layer of SiN as dielectric") == 0) &&
142  (strcmp(clientData,"capacitance") == 0)) {
```

```
143        dependentParmData = pcb_set_mks(dependentParmData, "overlap",
               1e-6*round(1e6*((capacitance-(112.5*Width +0.00225)*1e-12)
               * 0.3e-6)/(Width*7.5*8.85e-12)));
144        overlap=1e-6*round(1e6*((capacitance-(112.5*Width +0.00225)*1e
               -12)* 0.3e-6)/(Width*7.5*8.85e-12));
145    } else if (( strcmp(mod, "This structure consists of a 300nm thick
146    layer of AlN plus a 300nm thick layer of SiN on top as dielectric"
           )
147    == 0) && ((strcmp(clientData,"Width") == 0)||
148    (strcmp(clientData,"Structure") == 0))) {
149        dependentParmData = pcb_set_mks(dependentParmData, "overlap",
               1e-6*round(1e6*((capacitance-(112.5*Width +0.00225)*1e-12)
               * 0.3e-6)/(Width*4.286*8.85e-12)));
150        overlap=1e-6*round(1e6*((capacitance-(112.5*Width +0.00225)*1e
               -12)* 0.3e-6)/(Width*4.286*8.85e-12));
151    } else if (( strcmp(mod, "This structure consists of a 300nm thick
152    layer of AlN plus a 300nm thick layer of SiN on top as dielectric"
           )
153    == 0) && (strcmp(clientData,"overlap") == 0))
154
155        dependentParmData = pcb_set_mks(dependentParmData, "
               capacitance", 1e-14*round(1e14*(((overlap*Width*4.286*8.85
               e-12)/0.3e-6)+(112.5*Width +0.00225)*1e-12)));
156
157    else if (( strcmp(mod, "This structure consists of a 300nm thick
158    layer of AlN plus a 300nm thick layer of SiN on top as dielectric"
           )
159    == 0) && (strcmp(clientData,"capacitance") == 0)) {
160        dependentParmData = pcb_set_mks(dependentParmData, "overlap",
               1e-6*round(1e6*((capacitance-(112.5*Width +0.00225)*1e-12)
               * 0.3e-6)/(Width*4.286*8.85e-12)));
161        overlap=1e-6*round(1e6*((capacitance-(112.5*Width +0.00225)*1e
               -12)* 0.3e-6)/(Width*4.286*8.85e-12));
162    }
163    //if (strcmp(clientData,"B") == 0)
164    //dependentParmData = pcb_set_mks(dependentParmData, "GG", 1e-6*
           round(1e6*B/4));
165    //if (strcmp(clientData,"GG") == 0)
166    //dependentParmData = pcb_set_mks(dependentParmData, "B", 1e-6*
           round(1e6*GG*4));
167    if ( strcmp(mod, "This structure consists of a 300nm thick layer
           of
168    AlN as dielectric") == 0) dependentParmData =
169    pcb_set_mks(dependentParmData, "t", 1); else if ( strcmp(mod, "
           This
170    structure consists of a 300nm thick layer of SiN as dielectric")
           ==
171    0) dependentParmData = pcb_set_mks(dependentParmData, "t", 2);
           else
172    if ( strcmp(mod, "This structure consists of a 300nm thick layer
           of
173    AlN plus a 300nm thick layer of SiN on top as dielectric") == 0)
174    dependentParmData = pcb_set_mks(dependentParmData, "t", 3);
175
176    if(((strcmp(clientData,"overlap") == 0) ||
177    (strcmp(clientData,"Width") == 0) ||
178    (strcmp(clientData,"capacitance") == 0) ||
179    (strcmp(clientData,"Structure") == 0)) && (overlap < 20 um))
```

```
180  de_info("The model for the simulation will not be valid for
          overlaps
181  less than 20 um", 0); return dependentParmData; }
182  // here the part for the DC-block is finished
```

File name: dc_block_model

File Position: ISiT_design_kit/circuit/models

The following is the model file for the DC-block. This file contains three similar parts for the three different DC-blocks having three possible different dielectric combinations; the difference between these parts is only the formula for calculating the series capacitor in the model (here the comments appear after ; sign).

```
1   define ALN_STRUCTURE_SECTION ( n1 n2) parameters    Width=   GG=
2   overlap=   B=   capacitance=   t=
3
4   a=(Width*1000000/2)  b=(GG*1000000/2)  c=(B*1000000/2)  tt=b-a
5   q1=1-((b*b)/(c*c))  q2=1-((a*a)/(c*c))  q3=sqrt(q1/q2)  k=(a/b)*q3
6   k1=sqrt(1-pow(k,2))  elli= PI/(ln(2*(1+sqrt(k1))/(1-sqrt(k1))))
7   ;There is no need to have two different elliptic ratio functions
          for
8   different values of k ,because ;for k>0.707 the two different
9   elliptic ratio functions give values very close to each other
10  C31=(4*8.854e-12)*elli thicknessc=((4*8.854*1.5)/tt)*1e-12
11  L1=1/(9e16*(C31+thicknessc))
12  s1=1-((pow((sinh(PI*b/(2*508))),2))/(pow((sinh(PI*c/(2*508))),2)))
13  s2=1-((pow((sinh(PI*a/(2*508))),2))/(pow((sinh(PI*c/(2*508))),2)))
14  k6=((sinh((PI*a)/(2*508)))/(sinh((PI*b)/(2*508))))*sqrt(s1/s2)
15  k61=sqrt(1-k6*k6) ellic= PI/(ln(2*(1+sqrt(k61))/(1-sqrt(k61))))
16  C32=(2*8.854e-12*10.9)*ellic C2=(C31/2)*overlap*1e15 ;this is part
17  of the capacitance to the ground ;from the gold part which is
18  directly on the underpath
19   C3=thicknessc*(25e-6+overlap)*1e15
20  ;this is the capacitance due to ;thickness for the whole gold side
21  CUPPER=C31/2 CLOWER=(C31/2)+C32 w=2*a
22  delta=(1.25*3/PI)*(1+ln(4*PI*w/3))  w1=delta+w
23  weff=w1+(4/3.14)+ln(6.28*((w1/4)+0.92))  COXSMALL=2*8.85*weff*1e-12
24  rr=pow((tt-10),0.55)  ww=w-30+1e-13  rad=sqrt(ww)
25  COXBIG=(1200+79.1*rad+((110.6+12.03*rad)*rr))*1e-12
26  CSUB1=(CLOWER*COXSMALL)/(CLOWER+COXSMALL)
27  CSUB2=(CSUB1*COXBIG)/(CSUB1+COXBIG) C1=(CSUB2+C31/2)*25e-6*1e15
28
29  ;the total capacitance to the ground for the ;25um length of line
          on
30  the left(minus the thickness capacitance)
31
32  CP1VALUE=(C1+C2+C3) ;this is in correct unit
33
34  LVALUE=L1*(60e-6+overlap) ;correct unit
35
36  LVALUELEFT=LVALUE*((25e-6+overlap/2)/(overlap+60e-6))
37
38  LVALUERIGHT=LVALUE*((35e-6+overlap/2)/(overlap+60e-6))
39
40  fringmodel= 0.0001125*Width*1e6 + 0.00225 ;this will be in pF
41
42  CSVALUE=((8.85e-12*10*Width*1e6*overlap/0.3)*1e12+fringmodel)*1e
          -12
43  ;correct unit
```

```
44
45  CA2=CSUB2*overlap*1e15 ;the capacitance from overlaped undepath to
46  the ;ground
47
48  CA3=thicknessc*25e-6*1e15 ;notice that we have thick metal just
       for
49  25 um
50
51  CA1=(CSUB2+C31/2)*35e-6*1e15 ;the total capacitance to the ground
52  for the ;35um length of line on the right(minus the thickness
53  capacitance)
54
55  CP2VALUE=(CA1+CA2+CA3) ;this is in fF
56
57  Periph=Width*1000000+6 IWidth=1/Periph RS0=(0.0420*overlap
       *1000000+
58  2.4299)*IWidth+0.0148 fg0f=407.4861*IWidth+6.9491
59  Rs=RS0*sqrt(1+freq/(fg0f*1e9)) RVALUE=Rs*1.205
60
61  L:LSL n1 4 L=LVALUELEFT
62
63  C:CS 4 5 C=CSVALUE
64
65  Y_Port:Y2P1  5 0 6 0 Y[1,1]=1/RVALUE Y[1,2]=-1/RVALUE
66  Y[2,1]=-1/RVALUE Y[2,2]=1/RVALUE Recip=no
67
68  L:LSR 6 n2 L=LVALUERIGHT
69
70  C:CP1 4 0 C=CP1VALUE fF
71
72  C:CP2 6 0 C=CP2VALUE fF
73
74  end ALN_STRUCTURE_SECTION
75
76  define SiN_STRUCTURE_SECTION ( n1 n2) parameters   Width=  GG=
77  overlap=  B=  capacitance=  t=
78
79  a=(Width*1000000/2) b=(GG*1000000/2) c=(B*1000000/2) tt=b-a
80  q1=1-((b*b)/(c*c)) q2=1-((a*a)/(c*c)) q3=sqrt(q1/q2) k=(a/b)*q3
81  k1=sqrt(1-pow(k,2)) elli= PI/(ln(2*(1+sqrt(k1))/(1-sqrt(k1))))
82  C31=(4*8.854e-12)*elli thicknessc=((4*8.854*1.5)/tt)*1e-12
83  L1=1/(9e16*(C31+thicknessc))
84
85  s1=1-((pow((sinh(PI*b/(2*508))),2))/(pow((sinh(PI*c/(2*508))),2)))
86  s2=1-((pow((sinh(PI*a/(2*508))),2))/(pow((sinh(PI*c/(2*508))),2)))
87  k6=((sinh((PI*a)/(2*508)))/(sinh((PI*b)/(2*508))))*sqrt(s1/s2)
88  k61=sqrt(1-k6*k6) ellic= PI/(ln(2*(1+sqrt(k61))/(1-sqrt(k61))))
89  C32=(2*8.854e-12*10.9)*ellic C2=(C31/2)*overlap*1e15 ;this is part
90  of the capacitance to the ground ;from the gold part which is
91  directly on the underpath
92   C3=thicknessc*(25e-6+overlap)*1e15
93  ;this is the capacitance due to ;thickness for the whole gold side
94  CUPPER=C31/2 CLOWER=(C31/2)+C32 w=2*a
95  delta=(1.25*3/PI)*(1+ln(4*PI*w/3)) w1=delta+w
96  weff=w1+(4/3.14)+ln(6.28*((w1/4)+0.92)) COXSMALL=2*8.85*weff*1e-12
97  rr=pow((tt-10),0.55) ww=w-30+1e-13 rad=sqrt(ww)
98  COXBIG=(1200+79.1*rad+((110.6+12.03*rad)*rr))*1e-12
99  CSUB1=(CLOWER*COXSMALL)/(CLOWER+COXSMALL)
100 CSUB2=(CSUB1*COXBIG)/(CSUB1+COXBIG) C1=(CSUB2+C31/2)*25e-6*1e15
```

```
101
102  ;the total capacitance to the ground for the ;25um length of line
       on
103  the left(minus the thickness capacitance)
104
105  CP1VALUE=(C1+C2+C3) ;this is in correct unit
106
107  LVALUE=L1*(60e-6+overlap) ;correct unit
108
109  LVALUELEFT=LVALUE*((25e-6+overlap/2)/(overlap+60e-6))
110
111  LVALUERIGHT=LVALUE*((35e-6+overlap/2)/(overlap+60e-6))
112
113  fringmodel= 0.0001125*Width*1e6 + 0.00225 ;this will be in pF
114
115  CSVALUE=((8.85e-12*7.5*Width*1e6*overlap/0.3)*1e12+fringmodel)*1e
       -12
116  ;correct unit
117
118  CA2=CSUB2*overlap*1e15 ;the capacitance from overlaped undepath to
119  the ;ground
120
121  CA3=thicknessc*25e-6*1e15 ;notice that we have thick metal just
       for
122  25 um
123
124  CA1=(CSUB2+C31/2)*35e-6*1e15 ;the total capacitance to the ground
125  for the ;35um length of line on the right(minus the thickness
126  capacitance)
127
128  CP2VALUE=(CA1+CA2+CA3) ;this is in fF
129
130  Periph=Width*1000000+6 IWidth=1/Periph RS0=(0.0420*overlap
       *1000000+
131  2.4299)*IWidth+0.0148 fg0f=407.4861*IWidth+6.9491
132  Rs=RS0*sqrt(1+freq/(fg0f*1e9)) RVALUE=Rs*1.205
133
134  L:LSL n1 4 L=LVALUELEFT
135
136  C:CS 4 5 C=CSVALUE
137
138  Y_Port:Y2P1  5 0 6 0 Y[1,1]=1/RVALUE Y[1,2]=-1/RVALUE
139  Y[2,1]=-1/RVALUE Y[2,2]=1/RVALUE Recip=no
140
141  L:LSR 6 n2 L=LVALUERIGHT
142
143  C:CP1 4 0 C=CP1VALUE fF
144
145  C:CP2 6 0 C=CP2VALUE fF
146
147  end SiN_STRUCTURE_SECTION
148
149  define TWOLAYER_STRUCTURE_SECTION  ( n1 n2) parameters   Width=
       GG=
150  overlap= B= capacitance= t=
151
152  a=(Width*1000000/2) b=(GG*1000000/2) c=(B*1000000/2) tt=b-a
153  q1=1-((b*b)/(c*c)) q2=1-((a*a)/(c*c)) q3=sqrt(q1/q2) k=(a/b)*q3
154  k1=sqrt(1-pow(k,2)) elli= PI/(ln(2*(1+sqrt(k1))/(1-sqrt(k1))))
```

```
155  C31=(4*8.854e-12)*elli thicknessc=((4*8.854*1.5)/tt)*1e-12
156  L1=1/(9e16*(C31+thicknessc))
157  s1=1-((pow((sinh(PI*b/(2*508))),2))/(pow((sinh(PI*c/(2*508))),2)))
158  s2=1-((pow((sinh(PI*a/(2*508))),2))/(pow((sinh(PI*c/(2*508))),2)))
159  k6=((sinh((PI*a)/(2*508)))/(sinh((PI*b)/(2*508))))*sqrt(s1/s2)
160  k61=sqrt(1-k6*k6) ellic= PI/(ln(2*(1+sqrt(k61))/(1-sqrt(k61))))
161  C32=(2*8.854e-12*10.9)*ellic C2=(C31/2)*overlap*1e15 ;this is part
162  of the capacitance to the ground ;from the gold part which is
163  directly on the underpath
164    C3=thicknessc*(25e-6+overlap)*1e15
165  ;this is the capacitance due to ;thickness for the whole gold side
166  CUPPER=C31/2 CLOWER=(C31/2)+C32 w=2*a
167  delta=(1.25*3/PI)*(1+ln(4*PI*w/3)) w1=delta+w
168  weff=w1+(4/3.14)+ln(6.28*((w1/4)+0.92)) COXSMALL=2*8.85*weff*1e-12
169  rr=pow((tt-10),0.55) ww=w-30+1e-13 rad=sqrt(ww)
170  COXBIG=(1200+79.1*rad+((110.6+12.03*rad)*rr))*1e-12
171  CSUB1=(CLOWER*COXSMALL)/(CLOWER+COXSMALL)
172  CSUB2=(CSUB1*COXBIG)/(CSUB1+COXBIG) C1=(CSUB2+C31/2)*25e-6*1e15
173
174  ;the total capacitance to the ground for the ;25um length of line
       on
175  the left(minus the thickness capacitance)
176
177  CP1VALUE=(C1+C2+C3) ;this is in correct unit
178
179  LVALUE=L1*(60e-6+overlap) ;correct unit
180
181  LVALUELEFT=LVALUE*((25e-6+overlap/2)/(overlap+60e-6))
182
183  LVALUERIGHT=LVALUE*((35e-6+overlap/2)/(overlap+60e-6))
184
185  fringmodel= 0.0001125*Width*1e6 + 0.00225 ;this will be in pF
186
187  CSVALUE=((8.85e-12*4.2857*Width*1e6*overlap/0.3)*1e12+fringmodel)
       *1e-12
188  ;correct unit
189
190  CA2=CSUB2*overlap*1e15 ;the capacitance from overlaped undepath to
191  the ;ground
192
193  CA3=thicknessc*25e-6*1e15 ;notice that we have thick metal just
       for
194  25 um
195
196  CA1=(CSUB2+C31/2)*35e-6*1e15 ;the total capacitance to the ground
197  for the ;35um length of line on the right(minus the thickness
198  capacitance)
199
200  CP2VALUE=(CA1+CA2+CA3) ;this is in fF
201
202  Periph=Width*1000000+6 IWidth=1/Periph RS0=(0.0420*overlap
       *1000000+
203  2.4299)*IWidth+0.0148 fg0f=407.4861*IWidth+6.9491
204  Rs=RS0*sqrt(1+freq/(fg0f*1e9)) RVALUE=Rs*1.205
205
206  L:LSL n1 4 L=LVALUELEFT
207
208  C:CS 4 5 C=CSVALUE
209
```

```
210  Y_Port:Y2P1   5  0  6  0  Y[1,1]=1/RVALUE  Y[1,2]=-1/RVALUE
211  Y[2,1]=-1/RVALUE  Y[2,2]=1/RVALUE  Recip=no
212
213  L:LSR  6  n2  L=LVALUERIGHT
214
215  C:CP1  4  0  C=CP1VALUE  fF
216
217  C:CP2  6  0  C=CP2VALUE  fF
218
219  end  TWOLAYER_STRUCTURE_SECTION
```

File name: Isit_Inductorskit_models
File Position: ISiT_design_kit/circuit/models
The following file contains the simulation information for the inductors:

```
1   //This file contains the simulation information for the inductors
2
3   define Isit_Inductors_L0324 ( n1 n2) Options ResourceUsage=yes
4   UseNutmegFormat=no #uselib "ckt" , "S2P" S2P:SNP1   n1 n2 0
5   File="_24oneturnfinal_mom_a.ds" Type="dataset" Block="data"
6   InterpMode="linear" InterpDom="" ExtrapMode="constant" Temp=27.0
7   CheckPassivity=0 end Isit_Inductors_L0324
8
9   define Isit_Inductors_L05324 ( n1 n2) Options ResourceUsage=yes
10  UseNutmegFormat=no #uselib "ckt" , "S2P" S2P:SNP1   n1 n2 0
11  File="_1_5_60_20_10final_mom_a.ds" Type="dataset" Block="data"
12  InterpMode="linear" InterpDom="" ExtrapMode="constant" Temp=27.0
13  CheckPassivity=0 end Isit_Inductors_L05324
14
15  define Isit_Inductors_L08524 ( n1 n2) Options ResourceUsage=yes
16  UseNutmegFormat=no #uselib "ckt" , "S2P" S2P:SNP1   n1 n2 0
17  File="_1_5_110_20_10final_mom_a.ds" Type="dataset" Block="data"
18  InterpMode="linear" InterpDom="" ExtrapMode="constant" Temp=27.0
19  CheckPassivity=0 end Isit_Inductors_L08524
20
21  define Isit_Inductors_L124 ( n1 n2) Options ResourceUsage=yes
22  UseNutmegFormat=no #uselib "ckt" , "S2P" S2P:SNP1   n1 n2 0
23  File="_1_5_140_20_10final_mom_a.ds" Type="dataset" Block="data"
24  InterpMode="linear" InterpDom="" ExtrapMode="constant" Temp=27.0
25  CheckPassivity=0 end Isit_Inductors_L124
26
27  define Isit_Inductors_L1324 ( n1 n2) Options ResourceUsage=yes
28  UseNutmegFormat=no #uselib "ckt" , "S2P" S2P:SNP1   n1 n2 0
29  File="_2_5_70_20_10final_mom_a.ds" Type="dataset" Block="data"
30  InterpMode="linear" InterpDom="" ExtrapMode="constant" Temp=27.0
31  CheckPassivity=0 end Isit_Inductors_L1324
32
33  define Isit_Inductors_L0735 ( n1 n2) Options ResourceUsage=yes
34  UseNutmegFormat=no #uselib "ckt" , "S2P" S2P:SNP1   n1 n2 0
35  File="_1_5_90_20_10final_mom_a.ds" Type="dataset" Block="data"
36  InterpMode="linear" InterpDom="" ExtrapMode="constant" Temp=27.0
37  CheckPassivity=0 end Isit_Inductors_L0735
38
39
40  define Isit_Inductors_L0535 ( n1 n2) Options ResourceUsage=yes
41  UseNutmegFormat=no #uselib "ckt" , "S2P" S2P:SNP1   n1 n2 0
42  File="_1_5_40_20_20final_mom_a.ds" Type="dataset" Block="data"
43  InterpMode="linear" InterpDom="" ExtrapMode="constant" Temp=27.0
44  CheckPassivity=0 end Isit_Inductors_L0535
```

```
45
46   define Isit_Inductors_L07635 ( n1 n2) Options ResourceUsage=yes
47   UseNutmegFormat=no #uselib "ckt" , "S2P" S2P:SNP1   n1 n2  0
48   File="_2_5_40_10_10nickelfinal_mom_a.ds" Type="dataset" Block="
        data"
49   InterpMode="linear" InterpDom="" ExtrapMode="constant" Temp=27.0
50   CheckPassivity=0 end Isit_Inductors_L07635
51
52
53   define Isit_Inductors_L0335 ( n1 n2) Options ResourceUsage=yes
54   UseNutmegFormat=no #uselib "ckt" , "S2P" S2P:SNP1   n1 n2  0
55   File="_24oneturnfinal_mom_a.ds" Type="dataset" Block="data"
56   InterpMode="linear" InterpDom="" ExtrapMode="constant" Temp=27.0
57   CheckPassivity=0 end Isit_Inductors_L0335
```

References

1. Daniel L, Sangiovanni-Vincentelli A, White J (2002) Proximity templates for modeling of skin and proximity effects on packages and high frequency interconnect. In: Proceedings of the 2002 IEEE/ACM international conference on computer-aided design, ACM, New York, NY, USA, ICCAD '02, pp 326–333. doi:10.1145/774572.774621, http://doi.acm.org/10.1145/774572.774621
2. Gamble H, Armstrong B, Mitchell SJN, Wu Y, Fusco V, Stewart J (1999) Low-loss cpw lines on surface stabilized high-resistivity silicon. Microwave Guided Wave Lett IEEE 9(10):395–397. doi:10.1109/75.798027
3. Ghione G, Naldi CU (1987) Coplanar waveguides for mmic applications: effect of upper shielding, conductor backing, finite-extent ground planes, and line-to-line coupling. IEEE Trans Microwave Theory Tech 35(3):260–267. doi:10.1109/TMTT.1987.1133637
4. Gupta K, Garg R, Bahl I (1979) Microstrip lines and slotlines. Artech House, Incorporated
5. Horng TS, Peng K, Jau JK, Tsai YS (2003) S-parameter formulation of quality factor for a spiral inductor in generalized two-port configuration. In: Radio frequency integrated circuits (RFIC) symposium, 2003 IEEE, pp 255–258. doi:10.1109/RFIC.2003.1213938
6. Lee KY, Mohammadi S, Bhattacharya P, Katehi L (2006) Compact models based on transmission-line concept for integrated capacitors and inductors. IEEE Trans Microwave Theory Tech 54(12):4141–4148. doi:10.1109/TMTT.2006.886157
7. Mohan SS, del Mar Hershenson M (1999) Simple accurate expressions for planar spiral inductances. IEEE J Solid-State Circ 34(10):1419–1424. doi:10.1109/4.792620
8. Oh N, Lee S (2006) A simple model parameter extraction methodology for an on-chip spiral inductor. ETRI J 28(1):115–118
9. Pettenpaul E, Kapusta H, Weisgerber A, Mampe H, Luginsland J, Wolff I (1988) Cad models of lumped elements on gaas up to 18 ghz. IEEE Trans Microwave Theor Tech 36(2):294–304. doi:10.1109/22.3518
10. Sia CB, Ong BH, Chan KW, Yeo KS, Ma JG, Do A (2005) Physical layout design optimization of integrated spiral inductors for silicon-based rfic applications. IEEE Trans Electron Devices 52(12):2559–2567. doi:10.1109/TED.2005.859638
11. Valletta E, Van Beek J, Den Dekker A, Pulsford N, Jos HFF, De Vreede LCN, Nanver L, Burghartz J (2003) Design and characterization of integrated passive elements on high ohmic silicon. In: Microwave symposium digest, 2003 IEEE MTT-S international, vol 2, pp 1235–1238. doi:10.1109/MWSYM.2003.1212592
12. Wu Y, Gamble H, Armstrong B, Fusco V, Stewart J (1999) Sio2 interface layer effects on microwave loss of high-resistivity cpw line. Microwave Guided Wave Lett IEEE 9(1):10–12. doi:10.1109/75.752108
13. Yue C, Wong S (1998) On-chip spiral inductors with patterned ground shields for si-based rf ics. IEEE J Solid-State Circ 33(5):743–752. doi:10.1109/4.668989

© The Author(s) 2015
N. Pour Aryan, *Design and Modeling of Inductors, Capacitors and Coplanar Waveguides at Tens of GHz Frequencies*, SpringerBriefs in Electrical and Computer Engineering, DOI 10.1007/978-3-319-10187-3